Population Biology

AND

Evolution

Population Biology

AND

Evolution

Proceedings of the International Symposium
Sponsored by Syracuse University and the
New York State Science and Technology Foundation

JUNE 7–9, 1967

SYRACUSE, NEW YORK

Edited by

RICHARD C. LEWONTIN

Department of Biology
University of Chicago
Chicago, Illinois

SYRACUSE UNIVERSITY PRESS

First Edition

Library of Congress
Catalog Card Number: 68-30963

Printed in the United States of America

PREFACE

This book is the outcome of a Symposium on Population Biology held at Syracuse University on June 7, 8, and 9, 1967. What was originally conceived of as an intimate confrontation of geneticists, developmental biologists, and ecologists, soon grew to be a symposium of over 300 participants, including thirteen main speakers, a chairman, and uncounted commentators and questioners. So few of the latter produced a written version of their remarks that I have been forced, reluctantly, to leave out of this volume all trace of the spirited, and at times even aggressive, debate that went on. This is especially unfortunate because the fairly relaxed schedule of formal talks gave ample opportunity, an opportunity that was never wasted, for a prolonged discussion among the speakers and participants at large.

The intellectual successes of the Symposium were made possible by the strategic and tactical skill of many people. On the side of strategy, the Symposium owed its existence to Professor Frederick Sherman, Chairman of the Zoology Department at Syracuse University, whose idea it was in the first place, who invited me to come to Syracuse as Visiting Professor, in part to organize the Symposium, and whose gracious hospitality assured that everything would be easy. On the tactical side, Professor Marvin Druger of the Department of Zoology and Department of Science Education at Syracuse University so easily mastered the thousand problems of execution of a massive intellectual and social enterprise that he made it seem as if even I could have done it.

In the end, nothing could have been done without the very generous grant from the New York State Science and Technology Foundation, which supported the Symposium and made it possible to publish these proceedings.

R. C. Lewontin

Chicago, Fall, 1968

CONTENTS

Population Biology

AND

Evolution

Introduction

R. C. LEWONTIN
The University of Chicago
Chicago, Illinois

The twenty years since World War II have seen a vindication in biology of our faith in the Cartesian method as a way of doing science. Some of the most fundamental and interesting problems of biology have been solved or are very nearly solved by an analytic technique that is now loosely called "molecular biology." But it is not specifically the "molecular" aspect of the biology of the last twenty years that has led to its success. It is, rather, the analytic aspect, the belief that by breaking systems down into their component parts, by simplifying them or using simpler organisms, one can learn about more complex systems. As it happens, the problems that were attacked and are being attacked by this method lead to answers in terms of molecules and cell organelles.

And it is such terms that provide the only meaningful answers to the problem of gene action, subcellular morphogenesis, the nature of mutation, cellular physiology and metabolism, and the like. Moreover, it seems to me that molecular terms are the only meaningful ones in which to frame the answers to the question of development, although analytic methods have suffered some setbacks in developmental biology. This is also true of cytogenetics and the problems of crossing-over, chromosome pairing and chromosome movement although at the moment these are farthest from a satisfactory molecular description.

There is a host of problems in biology, however, that has been much neglected in these twenty exciting years, because the answers to them cannot be meaningfully framed in molecular and cellular terms. That is, the main trend of analytic biology, because it has become subcellular, has been away from a consideration of questions like: "What determines the rate of evolution of character?" "Why are there so many more species and so many fewer individuals of each species in the tropics than in the tundra?" "Why has sex appeared, disappeared, and taken on curious and aberrant forms in evolution?" Some of the most obvious and generally important questions in biology belong in this realm and remain unanswered. One of the most striking facts of life is the vast diversity of numbers and kinds of organisms. Why are some species world-wide yet rare and others restricted to a very small range yet tremendously abundant there? Why, in fact, are there different species at all? And, most curious of all, what is the significance of the fact that 99.999 per cent of all species are extinct? Now, such questions can be framed in molecular terms but not usefully, just as an apple can be described in quantum mechanical terms (in theory); but no one wants to know anything about an apple that a quantum mechanical description will tell him. Moreover, many important biological questions may remain unanswered precisely because the present trend of biology leads away from them. An example is the question, "Why is the primary sex-ratio in most animals one-to-one?" The answer at a mechanical-genetical level is framed in terms of chromosome disjunction and of the action of genes determining sex. But since, as I have already pointed out, these problems lead ultimately to molecular and subcellular investigations, the final answer of the genetical question is a molecular one. However, the peculiarity of this progression of investigation and analysis from chromosome to molecule is that it leads *away* from a completely different answer to the question "Why is the primary sex ratio one-to-one?" That different answer is an evolutionary one and takes into account the fact that chromosomal sex determination is not the only mechanism and that many organisms have other sex determining mechanisms which avoid the consequences of disjunction and meiosis and so have more flexible sex ratios. Different sex determination mechanisms result in different sex ratios, and it is a problem of population biology to discover under what circumstances a one-to-one sex ratio is evolved.

Unfortunately, both population biologists and cellular and molecular biologists have become confused about the differences between their modes of thought. It is *not* the case that molecular biology is Cartesian and analytic while population biology is holistic. Population biology is properly analytic and operates, within the framework of its own problems, by the process of simplification, analysis and resynthesis. It is unhappily true that there are population biologists who reject the analytic method and insist that the problems of ecology and evolution are so complex that they cannot be treated except by holistic statements. The influence of these people has held up progress in population biology for many years and, in addition, has tended to degrade population biology as a science. They are the stamp collectors of biology who, because they themselves are unable to analyze the complex problems of ecology and evolution, try to convince the rest of us that nothing but "objective description" of nature is possible.

Analytic population biology is attempting to deal with some of the most difficult problems of nature. It is proceeding by two paths. First, there is the experimental-inductive approach. Deliberately simplified systems of organisms interacting with a specified environment are set up in the laboratory in an attempt to dissect the complex interaction in nature. Recently, a number of us have started to expand the area of these experiments to include both genetic factors and ecological factors as variable, but always trying to keep as many variables constant as we can. I reject categorically the critics of this laboratory approach who keep saying "but Nature is different from the laboratory." Of course it is different. So what?

Second, there is the theoretical approach. There is again a wide misunderstanding of the function of theoretical studies in population biology. It is *not* the function of theory to describe what has happened in a particular instance. Only observation can do that. The purpose of theoretical studies in population biology is *to set limits*, to say, "This can happen; this cannot. This process will be extremely slow or of extremely small magnitude as compared to that process." Theoretical population biology is the science of the *possible*; only direct observation can yield a knowledge of the actual. But theoretical studies can then put limits on the experimental and observational procedures of observers and can also "explain" the results of experiments and observations.

If analytical population biologists are to solve the problems they have set themselves, they must be prepared to break out of old academic categories. A proper understanding of the interactions between species in a community, classically the province of ecology, is impossible without a consideration of the evolution and variation of the species themselves. Species are not static entities with fixed relations to the environment, but plastic elements, changing their genetic constitution under the influence of the physical factors of the environment and of the interactions with other species. The proper description of the interaction between, say, a predator and its prey, must include the fact that the physiological determinants of their interaction have evolved and are continuing to do so. No population ecology can succeed that is not also population genetics. On the other hand, the prediction of genetical changes in populations, the standard stuff of evolutionary genetics, is quite impossible without a specification of the relationship between phenotype and genotype in a fluctuating and uncertain environment. Different organisms, and different aspects of the same organisms, vary in their developmental sensitivity to external variations. Moreover, this developmental sensitivity is itself under the influence of genes and, thus, itself evolves. No population genetics can succeed that is not also a study of development. Finally, the modes of adaption of species to their environments, the ways they avoid extinction in an uncertain universe, include individual somatic plasticity, genetical variations within populations, and interspecific interaction in the community. Thus, the study of adaptation is the nexus of population ecology, population genetics and development. No one can pretend to an understanding of the history and natural economy of organisms, who does not include in his philosophy strains from all of these fields.

The planners of this symposium have attempted to bring together developmental biologists, ecologists, and geneticists who have shown in the past an awareness of the interdependence of their subjects. We hoped that they would feel a community of concerns and speak to each other in a common language. We succeeded only in part. Geneticists, embryologists, and ecologists remained untransmuted. But they did talk to each other, and the reader of this book cannot fail to see the mutualism that has grown, fostered in part by our symposium. There is not a paper in this volume that can be said to be strictly genetics or development or ecology. All are tainted, and the degree to which the languages and concepts of the different fields are mixed in them is a measure of the success of our enterprise.

-1-

The Spectrum of Genetic Variation

ALAN ROBERTSON
Institute of Animal Genetics,
West Mains Road, Edinburgh, Scotland

I would like to discuss from two rather different points of view the genetic variation in random breeding populations, particularly of animals, and to speculate a little on the difficulties of analyzing the natural selection process insofar as it affects genetic variation in continuously varying characters, which we cannot attribute to specific individual loci. I do not propose to discuss the mechanisms for the maintenance of the genetic variation, which I am sure will be dealt with by other speakers in the symposium.

We may observe genetic variation at two levels. Firstly, we may be in a position to identify genotypes and to discuss selection processes in terms of observable gene frequencies. I have in mind here particularly loci controlling antigenic differences and those controlling the specification of peptide chains for which analysis by starch gel electrophoresis is possible. In the last decade, mostly through the development of the starch gel technique, this field has expanded very rapidly. The second kind of genetic variation is that affecting continuously varying characters which is observed either by the response of the population to artificial selection, or by analyses of the similarities for the character between relatives of different kinds.

IDENTIFIABLE POLYMORPHISMS

We can reasonably ask three separate questions about the kind of genetic variation in which we can identify a high proportion of genotypes and make some estimate of gene frequencies.

1. What proportion of loci are segregating? To the classical geneticist, this is an unanswerable question because he only detects a locus by observ-

ing segregation at it. Nevertheless, we can now ask the question sensibly—at least of those cistrons concerned with the specification of a peptide chain. The peptide, or a larger molecule of which it forms a part, may then be identifiable on a starch gel, and we can then ask what proportion of such bands show evidence of segregation. Clearly, we are in some slight difficulty with those bands which are in fact found to be invariant, because we cannot be sure that all the peptides concerned are produced by different cistrons. We do, in fact, make the assumption that, if the biochemical activities of the peptides are completely different they are produced by different cistrons.

Recently, two investigations have been carried out to answer this question. The first, by Harris [1], dealt with enzymes in blood serum or in placental material in humans. Whether or not a particular enzyme could be examined depended mostly on the possibility of developing a technique for its identification on a starch gel. Ten enzymes were chosen, with no *a priori* reason to suspect polymorphism, and this was in fact found in three of them. More recently, working with *Drosophila pseudoobscura*, Lewontin and Hubby [2] have examined 21 different systems and discovered polymorphism in eight of them. Again there was no *a priori* reason to suspect that these particular peptides were polymorphic. The material is still rather sparse, but the indications are very strong that there is segregation at a high proportion of cistrons specifying peptide chains. The observed proportion must be an underestimate in that the experimental methods used depend on the existence of an electrical charge difference between the variants. It has been calculated that only about 27 per cent of single base substitutions conpatible with the genetic code will cause a change in the electric charge between the relevant amino acids. The extent to which the proportion of segregating loci is underestimated will then depend on the number of variants which are present at reasonable frequency at each locus in the population, and on the nature of the difference between them. If these are due to a single substitution, then, if there are only two alternatives at high frequency, we might expect to detect only 27 per cent of the polymorphic loci. However, if there are more alternative alleles in the population, our chance of observing a charge difference between at least two becomes higher. If, for instance, there are five and the sample size is large enough that they will all be represented, the chance that the polymorphism will be detected is 0.72. Whatever the fraction of polymorphic loci detected by this method, we must certainly face a situation in which, in these species, thousands of loci must be segregating.

Of course we cannot be sure how representative this class of loci is. A proportion of the DNA must be concerned with the production of the RNA machinery required for protein synthesis, not to speak of that part concerned with regulation. However, the recent confirmation in at least one case that a repressor substance is a protein might suggest that the greater part of the DNA is concerned with the specification of amino acid sequences.

2. How many alternatives are there at these segregating loci? Here again we face the problem of our inability to detect all the alternatives; but, nevertheless, I would suggest that the weight of the evidence is that, at least at those cistrons specifying a peptide chain, the number of alternatives held at reasonable frequencies in populations is not large, (i.e., is less than ten) although there are one or two examples where more have been found. The classic case with a great many alternative alleles, the B blood group locus in cattle, is not evidence on this point, as quite clearly this is a multiple locus and more than ten crossovers at the locus have now been detected [3].

This has important consequences in light of the modern view of the nature of mutation. If the locus concerned specifies an amino acid sequence by means of a sequence of bases perhaps 500 long, some of the old concepts of a balance between back and forward mutation, for instance, appear to be quite unreal. Mutational steps will very rarely repeat themselves, and we are faced rather with a continuing sequence of mutational change. If a high proportion of the mutations at a locus were then almost neutral in their effect on fitness, we would expect to find one of two alternative situations. On the one hand, we might find that very many alternative alleles were segregating in large populations. If, on the other hand, the population had been split into more or less isolated subpopulations, we might then expect to find a small number of alternatives in such populations, but with a different set in each. In fact, we find neither of these situations. We find very few alternatives, and on the whole the same ones are segregating in different populations. In my view, this must mean that the great majority of mutations are harmful and, conversely, that the great majority of polymorphisms have, at some time, been actively maintained by selection, although this selection need no longer exist to maintain the genetic diversity. One might also speculate as to whether some of the classical polymorphisms in human populations such as sickle cell haemoglobin have arisen by repeated mutations at the same site in the molecule, or whether all the sickle cell alleles are in fact replicates of one original mutation. It must not be forgotten that some of the human populations at present showing sickle cell polymorphism in a malarial environment in Africa may not have been in their present habitat for very long; perhaps less than 50 generations.

We are, then, faced with the paradox that the vast majority of new alternatives available to us within a species may be inferior to the standard haemoglobin type, whereas at the same time across species we find that there are a considerable number of haemoglobins doing a perfectly good job of oxygen transport in their own species. Does this mean that the effect of a given amino acid substitution on the functional efficiency of the haemoglobin molecule depends very much on the background within which it occurs—whether this is of the rest of the haemoglobin molecule itself or other enzymatic and physiological processes within the animal?

3. What is the nature of the difference between the alternatives? As I have already mentioned, in the case of sickle cell haemoglobin, the difference between the two alternatives is known to be a single amino acid substitution. But recently several cases have come to light in which this is not so. Domestic cattle, for instance, show a widespread haemoglobin polymorphism with the same two alternatives, A and B, segregating in very many of the known breeds. It now appears that the difference between these two alternatives is entirely in the non-α-chain and that there are three substitutions widely scattered along the chain, so that they could hardly have been produced by a single mutational event [4]. This would imply either that the present cattle population derived from a cross between two populations between which there had been divergence in haemoglobin structure in the past, or else that we are dealing with a polymorphism which is very long lived and which has accumulated three amino acid differences over this period. If this is so, we have a situation which somewhat parallels the inversion situation in *Drosophila* but within a single cistron, i.e., the gradual accumulation of genetic differences between the two alternative orders. There is evidence from other polymorphisms affecting milk proteins in cattle that here again some of the alternatives may differ by several amino acids, but there is no evidence as yet on the sequence of these [5].

The occurrences of repeat loci—often very close to one another on the chromosome—are now well supported by evidence and could well be of very frequent occurrence in the hereditary material. More recently, a clear case has been found in the three milk caseins of cattle. There are three casein fractions with similar but distinct gross amino acid compositions which are clearly controlled by loci which are very close to one another on the chromosome. As a large-animal geneticist, I am delighted to see that domestic animals are now contributing to our knowledge of basic genetics [6].

GENETIC VARIATION IN CONTINUOUS CHARACTERS

There is plenty of evidence that much of the variation in characters which vary continuously is genetic in origin. This comes from two sources.

First, selection experiments on laboratory animals—Drosophila, Tribolium [7], mice [9], rats, rabbits and so on, have shown that the mean of many characters can be altered several-fold by continued selection. For instance, my own standard outbred population of Drosophila has a mean sternopleural bristle count of about 17. Selection, of the kind that we normally do in the laboratory (10/25 in each generation), can produce lines averaging 12 bristles in the downward direction and 30 bristles in the upward direction. If an attempt is made to extract all the useful variation from the base population, we can produce high lines with an average score of over 50 and low lines with an average score of 7. We can show by chromosomal manipulations that selection pressure is affecting loci on all four chromosomes. Similar examples could easily be given for selection for body size in Tribolium or in mice, etc., as well as from selection experiments on domestic animals. For instance, selection of a random breeding population of White Leghorn chickens [8] has, in as few as six generations produced lines differing by more than a factor of two in body weight, and, if selection is applied to egg weight, by a factor rather less than this. Selection applied very simply—with 6 males and 18 breeding females in each generation—to back fat thickness in pigs has produced strains differing by a factor of two in about the same time.

The other kind of evidence comes from a statistical analysis of the similarities between related individuals in measurable characters, which gives a measure of how much of the phenotypic variation is genetic in origin. Here we obtain surprisingly high results for many measurements, expecially those which are of trivial importance to the reproductive ability of the organism. We find that about half of the variation in bristle number in Drosophila or of protein content in cow's milk—to give two diverse examples—is genetic in origin. The high degree of similarity between identical twins, both in cattle and in human populations, is again evidence that genetic variation is not lacking.

What do we know about the nature of this variation? This can be described under several headings.

1. How many loci affect the variation in a particular measurement? This is not a very useful question to ask because we might expect that allele substitutions at different loci would vary a good deal in their effect on the character measured. I believe that the shape of the distribution of gene effects on any measurement will be such that we will find a large number of loci having rather small effects on the character, even though a high proportion of the variation can be accounted for by a small number of loci. In Thoday's department in Cambridge and in our group in Edinburgh we have attempted to identify the individual loci responsible for differences between extreme selected lines. There can be no doubt now that these loci

can be identified and located on the chromosome and that a high proportion of the differences between extreme lines can then be accounted for by a comparatively samll number of loci; perhaps not very much more than 10 [10].

2. What are the gene frequencies at these loci—indeed, how many alleles at there at each? We have no evidence at all on the last point. If we assume that there are not many alleles at each locus, I deduce from my own selection experiments that the alleles that are usually fixed probably have initial frequencies not far away from 0.5. Nevertheless, the fact that we can, by the crossing of selected lines and by special experiments to extract all useful variation from the initial population, go far beyond the usual limits indicates to me that there are useful alleles at low frequencies which we may have difficulty in picking up with the usual selection procedure.

3. What are the linkage relationships at these loci? Having only recently discovered any loci at all that affect bristle characters in Drosophila, it is not surprising that we know very little about the linkage relationships between them. My own view for sternopleural bristles in Drosophila, mostly based on selection experiments in which we have completely suppressed crossing-over in some lines, would be that linkage in this character is not too far away from equilibrium in our standard population. I would not, however, wish to generalize this as there is good evidence from other sources (for instance Sheppard's work on butterflies) that "supergenes" exist by means of which useful attributes are collected together to segregate as one unit.

4. What is the nature of the gene action at these loci? There is good evidence, though most of it of an indirect statistical kind, that in characters which bear little relationship to the reproductive fitness of the organisms, gene action must be more or less additive in the sense of intermediacy of heterozygotes. However, I should stress that this has not to any extent been confirmed by the manipulation of the individual loci concerned.

I should perhaps also stress that in my discussion of genetic variation in populations I have tended to talk about characters that are peripheral in the sense that they have little relevance for reproductive fitness. Of course, selection has been applied to characters more closely related to fitness and, though there is evidence for genetic variation, response to selection in a favorable direction is usually not very marked. These characters, in general, show a considerable depression upon inbreeding and the argument as to whether genetic variation here is due to overdominance or to recessives at low frequency held in the population by mutation still goes on. I, myself, would take an intermediate point of view.

It is often stressed that the property of dominance is a function of the character but not of the locus. This may not always be true. There are, of course, instances in which the dominance relationships between two alleles may be reversed, depending on the characters being observed. Indeed considering the agouti gene A and the tan gene a^t in rabbits, A is dominant on the back and a^t on the belly. But even with characteristics like sterno-pleural bristle count, it has been my experience in examining the effects of major genes, characterised primarily by their effects on other parts of the individual, that if they are recessive for their main effect they are also recessive for bristle count.

The general question of the kind of gene action which affects a particular character is, considered in its evolutionary aspects, an extremely wide one. We must believe that the gene action which we see at present is a consequence of natural selection in the past—that the relationship of the character to reproductive fitness in the past is the main determinant of the nature and amount of gene action which we observe at the present time. Nevertheless, I would suggest that, in spite of the considerable amount of thought given to the consequences of natural selection when observed at the phenotypic level, we still know far too little about the process itself. Indeed, I sometimes think that it is prejudging the issue to talk about natural selection "acting" at all.

THE ANALYSIS OF THE NATURAL SELECTION PROCESS

The process of natural selection may be discussed in terms of the relative selective advantages of the different genotypes. Although this is useful in the discussions of changes of gene frequency, it is dismissed by some geneticists as not at all the right kind of approach. It tells us nothing whatsoever about the way in which natural selection takes place at the phenotypic level. We can express the relationship between the effect of the genotype on the phenotype and its effect in natural selection in the following diagram.

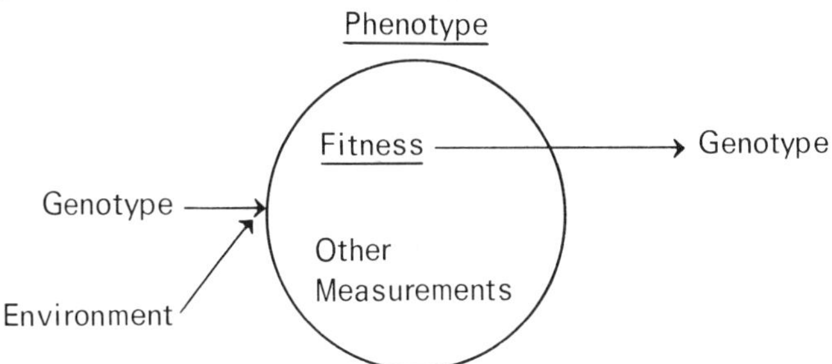

The relationship between genotype and reproductive fitness will then determine the changes in gene frequency which will take place as a result of the natural selection, but we are interested here in the relationship between genotype and phenotype (in the evolution of gene action) and in the relationship between phenotype and reproductive fitness. To be more precise, we are concerned with some particular phenotypic measurements. In this context I suggest that the word "phenotype" can only be used to mean the whole process of development, of which one particular aspect, though clearly the most important one, is the fitness; the number of gametes contributed to future generation. In general, we are dealing with the nature of the gene action on a particular measurement and with the relationship between the measurement and reproductive fitness. Such relationships are often discussed as though they summarised "natural selection acting on this particular measurement." In fact, we are observing the relationship between two particular aspects of the phenotype. We get into difficulties if we think of the value for the character as in some way determining the reproductive fitness.

Let me illustrate some of these, arising essentially from our inability to distinguish clearly cause and affect, by discussing the two quite different models which will lead to individuals with intermediate values for a particular measurement having superior reproductive fitness. I present two models merely as two extremes, rather than as the only two alternatives.

In the first we assume that the genetic variation is maintained in the population by heterozygotes having a superior reproductive fitness, for reasons other than the effects of the loci on the metric character for which we assume that there is additive gene action. If the gene frequencies are fairly close to 0.5, then individuals intermediate for the metric trait will have a higher proportion of loci heterozygous, and consequently a higher fitness, than will the extremes. This situation will be stable. The gene frequencies will remain constant as will the relationship between the character and fitness.

On the other hand there have been many papers published on the assumption that in such a relationship, the extreme individuals for the measurement are less fit because they are more extreme. This has been called "normalising selection" or "stabilising selection" and, as a model, "the optimum model." We should then expect there to be linkage disequilibrium between loci affecting the metric character and that, in the long run, the genetic variation will decline. Although we observe the same phenotypic relationship between the character and fitness in these two models, we see that the effect of the natural selection is quite different in the two. What is the solution to this difficulty?

Let me mention first two ways in which solutions are *not* to be found.

The first involves a kind of division of the organism into separate compartments, so that genes acting within a certain compartment only affect fitness through the end product of that compartment. This is hidden in the second model above. We are saying that the genes controlling this character alter fitness only through their effect on the character. This I would describe as a Berkeleyist view of the situation, i.e., that these loci affect fitness only through the particular character that I happen to be looking at. The second way around the problem—a little more sophisticated—is the *ceteris paribus* or "partial regression" approach. In this method we analyze the effect of variation in a particular measurement or characteristic with all others held equal. This is a nice trick for the statistician, but rather difficult for the animal itself to perform. When one asks "what other characteristics are we holding equal?" it becomes clear that this is really a completely abiological point of view, as indeed is also the compartmental approach.

Having disposed of two approaches which I do not regard as very valuable, it is incumbent on me to make some positive contribution to the situation. Let us be quite clear about the problem. We observe that individuals intermediate for a given measurement are fitter. We than ask what information we need to make predictions about the consequences of this relationship. How much of this "normalizing selection" really has a normalizing effect?

If we observe a selection process merely in terms of the relationship between various measurements and reproductive fitness, it is very difficult to make any predictions at all about the consequences, even about the mean of the measurements. This may come as a surprise to the quantitative geneticists, who are used to predicting genetic gain by multiplying the phenotypic selection differential by the heritability. This can only be done if it is known that the selection differential has come about because selection is applicable for that trait, and for that alone. If it is not known how the selection differential has come about, the formula cannot be used. In fact, one cannot, using the formula, even be sure of predicting the sign of the change in the mean of the population correctly. We know how to get around this problem in some cases by using what one might call the "secondary theorem of natural selection."

Fisher's fundamental theorem of natural selection may be stated rather crudely as the prediction that the change of reproductive fitness in any generation (or more accurately the Malthusian parameter) is equal to its additive genetic variance. The secondary theorem of natural selection states that the change in any character produced by a selection process is equal to the additive genetic covariance between fitness and the character itself. We may not always be able to observe this, but the formula has proved useful in the examination of some situations in animal breeding in which the fundamental nature of the selection decisions is far from clear. However, this does still not answer my basic question—how do I decide that the observed relationship between a phenotypic measurement and fitness is normalizing in consequences or not? I have to give the simple answer that, in general, I do not know how to do it.

I do know that by manipulating chromosomes in my Drosophila populations I can show that the changes of gene frequencies at loci affecting bristles on the third chromosome, under natural selection, are not dependent on the level of bristle score due to the rest of the genotype. The "optimum model" involves an interaction between loci in their effect on fitness. If it is correct, then if we make many loci homozygous for alleles with a positive effect on bristles, we should expect that natural selection acting on loci still segregating will reduce the bristle score by increasing the fitness of flies carrying the low alleles, and the reverse should happen if we make many loci homozygous for the low alleles. This I have done by examining the effect of the segregation of high and low third chromosomes in a background either of high chromosomes or low chromosomes respectively for the rest of the genotype. When the rest of the genetic background is low, natural selection does not increase bristle score, and the same is true in a high background.

I would suggest that this whole discussion might be looked upon as an illustration of the necessity to impute motive when we discuss the effect of natural selection on the phenotype as a whole. If we look at a particular property of an organism, how can we be sure that it has been selected into this population for itself alone, or that it is not merely the consequence of a selection process which had completely different ends at the time. Clearly, many feedback circuits in organisms have been deliberately selected for. However, we must not forget that feedback is a perfectly usual property of chemical systems. Many of you who did physical chemistry will remember, for instance, that Le Chatelier's principle states that any chemical system in equilibrium will respond to a change in the environment in such a way as to reduce the immediate effect of that change. Rather more generally, I suspect that we may find that many of the properties of organisms, which have been argued for as having been specially selected for, may turn out to be necessary consequences of the structure of complex organisms. For instance, if the flow through a series of sequential chemical reactions is examined, it is found that the system cannot be at the same time highly sensitive to changes in parameters affecting all of the reactions. A necessary consequence of the system (and this does not depend on the reactions being enzyme-controlled) is that the rate of flow through the system is extremely insensitive to changes in the parameters at the majority of the links in the chain. One has immediately a model for dominance at such loci. Our basic problem is, perhaps, that in a biological system we can never be sure of the level at which we can apply inductive argument. I would suggest that we have to be very clear about the nature of the entities we manipulate before we can do this. In discussing the way in which natural selection will operate on the complex physical and biochemical network which is an organism, we must distinguish clearly between at least three different classes of observation. These are:

(a) environmental variables;

(b) internal measurements, which we may in principle be able to make without the developmental process itself having to take place. (Clearly one of these is the genotype itself. If we know the genotype we may further be able to specify the properties of a large number of the biochemical molecules which will then be produced, and these are statements which we can make whether or not development takes place. We can, for instance, if we know that a particular individual is heterozygous for sickle cell haemoglobin, make certain valid predictions about the ability of his haemoglobin to react with oxygen at different pressures);

(c) measurements which we can only make in the course of the developmental process. These are the true observables—the other two classes of measurements are in a sense parameters which determine the course of development.

I would suggest, then, that we must be very clear in inductive argument so that we do not include as causes some observations in the third group that are really not causes at all, but effects. Unless we make this distinction in the discussion of the evolution of complex systems, we are likely to remain very confused.

LITERATURE CITED

1. Harris, H., 1966, "Enzyme Polymorphisms in Man." *Proc. Roy. Soc. London B*, 164:298–310.
2. Lewontin, R. C., and J. L. Hubby, 1966, "A Molecular Approach to the Study of Genic Heterozygosity in Natural Populations," II. "Amount of Variation and Degree of Heterozygosity in Natural Populations of *Drosophila pseudoobscura*," *Genetics*, 54:595–609.
3. Bouw, J., G. E. Nasrat, and C. Buys, 1964, "The Inheritance of Blood Groups in the Blood Group System B in Cattle," *Genetica*, 35:47–58.
4. Schroeder, W. A., and R. T. Jones, 1965, "Some Aspects of the Chemistry and Function of Human and Animal Hemoglobins," *Fortscher. Chem. Org. Naturstoffe*, 23:113–194.
5. Gordon, W. G., J. J. Basch, and M. P. Thompson, 1965, "Genetic Polymorphism in

Caseins of Cow's Milk," VI. Amino Acid Composition of α_{s1}-Caseins of A, B, and C." *J. Dairy Sci.*, 48:1010–1018.

6. King, J. W. B., R. Aschaffenburg, C. A. Kiddy, and M. P. Thompson, 1965, "Non-independent Occurrence of α and β-casein Variants of Cow's Milk," *Nature* 206:324–325.
7. Hardin, R. T., and A. E. Bell, 1967, "Two-way Selection for Body Weight in Tribolium on Two Levels of Nutrition," *Genet. Rec.*, 9:309–330.
8. Festing, M. F., and A. W. Nordskog, 1967, "Response to Selection for Body Weight and Egg Weight in Chickens." *Genetics*, 55:219–231.
9. Falconer, D. S., and M. Latyszewski, 1952, "The Environment in Relation to Selection for Size in Mice," *J. Genet.*, 51:67–80.
10. Thoday, J. M., J. B. Gibson, and S. G. Spickett, 1964, "Regular Responses to Selection. 2. Recombination and Accelerated Response," *Genet. Res.*, 5:1–19.

- 2 -

Integration of Development and Evolutionary Progress

G. LEDYARD STEBBINS
University of California
Davis, California

The most important feature of the modern synthetic theory of evolution is its foundation upon a great variety of biological disciplines. Nevertheless, the contributions made by various fields of biology have been very unequal. Some of this inequality is due to the fact that certain disciplines, such as population genetics and paleontology, can by their very nature provide a disproportionate share of information which is valuable to evolutionary theory. On the other hand, other disciplines have as yet contributed far less than their appropriate and potential share of facts. One of these is developmental biology.

Until relatively recent times, developmental biology or embryology was associated with evolutionary theory largely by means of the recapitulation theory of Haeckel, by von Baer's theory of embryonic similarity, and by related variants of the general idea that past lines of phylogeny can be revealed through studying the development of modern forms. The difficulties of such studies, and the mistaken concepts to which they can lead, have been pointed out by a number of authors. Because of them, the embryological approach to phylogeny is now much less fashionable than it was in the past. The newer, experimental embryology has up to now concerned itself so much with mechanisms of development in a few "classical" examples of animals and plants that it has gravely neglected the comparative aspects of the field. For this reason, the contributions of embryology to evolutionary theory have declined considerably during the past half century.

A revival of interest in the application of embryology to evolutionary theory has come through a better understanding of developmental genetics, that is, the pathway from the gene to the character. This avenue has been opened up largely by the syntheses of facts and the theoretical models developed by Waddington (1962). Contributions from this direction, contrary to the older embryology, are helping us understand not particular lines of phylogeny, but the operation of universal evolutionary processes, particularly mutation and selection. This is the relationship between developmental biology and evolution which I wish to discuss here.

DEVELOPMENTAL PATHWAYS AND NATURAL SELECTION

Failure to appreciate the complexity of gene action in development has been responsible for much confused thinking about the action of natural selection. The problem lies in the fact that the selective action of the external environment, by its very nature, can operate only upon individual phenotypes, through their survival, death, or differential rate of reproduction. Evolution, however, must be measured not in terms of altered phenotypes, but as a function of changes in the composition of the gene pool. Between the act of selection and the response of the population lie all the complexities of gene action in development.

These complexities are exhibited largely in terms of the pleiotropic action of genes. While the fact is now firmly established that a particular gene codes for only one polypeptide chain, which may form either an enzyme, a part of an enzyme molecule, or a molecule of structural protein, these primary results of gene action are only of indirect importance to selection. In any multicellular animal or plant, the product of each enzyme-controlled reaction interacts in various ways with the products of other reactions, so that if any such reaction is altered by mutation, the effects on the ultimate phenotype are inevitably numerous and diverse. Examples of such secondary pleiotropism are well known even to beginning students of genetics.

The most important consequence of this secondary pleiotropism is that selection for any characteristic is certain to alter other characteristics of the organism for which no direct selection is being practiced. For instance, Lerner (1954) has shown that selection for increased shank length in chickens alters the entire conformation of the animal. In our laboratory, we are now analyzing an example in higher plants which shows that these side effects can vary greatly from one selected individual or line to another. In an annual plant species native to California, *Linanthus androsaceus*, the corollas are normally five-lobed, but plants with six- or seven-lobed corollas occur at low frequencies in nearly every population. Huether (1966) found that five generations of selection for high corolla lobe number produced a more than five-fold increase in the frequency of this condition; and in some plants of his selected lines, corollas with as many as 16 lobes could be found. Analyzing the flowers of some progeny of these selected lines, we have found that the other organs of the flower are variously affected. In most plants, the normal number of five stamens per flower has been retained, while in a few of them, flowers containing six or more stamens are common. In several plants containing extra corolla lobes, the number of stamens is reduced, the expected position of some stamens being occupied by extra corolla lobes. For the most part, the ovaries of these selected plants have the expected number of three carpels bearing three stigmas; but ovaries with extra numbers of stigmas are occasionally found, and in one plant the mean number of stigmas was as high as 5.5.

An analysis of the developmental basis of these abnormalities has just begun. Preliminary results suggest that plants with extra numbers of parts have floral meristems with larger numbers of cells at the time of primordium differentiation, presumably because of an increased frequency of mitoses at these stages. The differences between individuals may well be due to the fact that genes can bring about altered numbers of corolla lobes by changing the frequency of mitoses at any one of several stages in the development of floral primordia.

In various plant groups there is good evidence that both artificial and natural selection for a generalized adaptive characteristic, size of fruits or seeds, has produced as a nonadaptive by-product alterations in the numbers of parts of the flower. A well-known example is the cultivated tomato, *Lycopersicum esculentum*. In all of the wild species of its genus, as well as of the related and much larger genus *Solanum*, the flowers have five calyx lobes, five corolla lobes, five stamens, and a two-loculate ovary bearing two stigmas. This condition exists in the smaller fruited varieties of the cultivated species. The large fruited cultivars of tomato, however, usually have six or seven calyx and corolla lobes, and one or two extra, usually imperfectly developed, ovary locules. The development of the flower in different varieties of tomato was studied many years ago in the laboratory of E. W. Sinnott, by Houghtaling (1935). She found that even before the calyx and corolla are differentiated, the floral meristem of the large fruited varieties is larger and contains more cells than does the corresponding meristem of the small fruited varieties. Apparently, some of the selected genes or gene combinations favoring increased fruit size accomplish this result by increasing the size of the very young floral meristem. If we assume that the number of undifferentiated meristematic cells required for the formation of a calyx or corolla lobe primordium is relatively constant, then any increase in the overall number of cells in the meristem will inevitably increase the number of floral parts. Comparable increases in the number of floral parts are found in large fruited cultivars of other species, such as the eggplant (*Solanum melongena*), pepper (*Capsicum annuum*), and pomegranate (*Punica granatum*).

Comparative studies of wild plant species indicate that natural selection for increased fruit, seed, or flower size has produced similar adaptively neutral alterations in the numbers of floral parts. An example from the California flora is a shrub, *Ceanothus Jepsonii*, endemic to dry serpentine soils in the north Coast Ranges. In all other species of the large genus

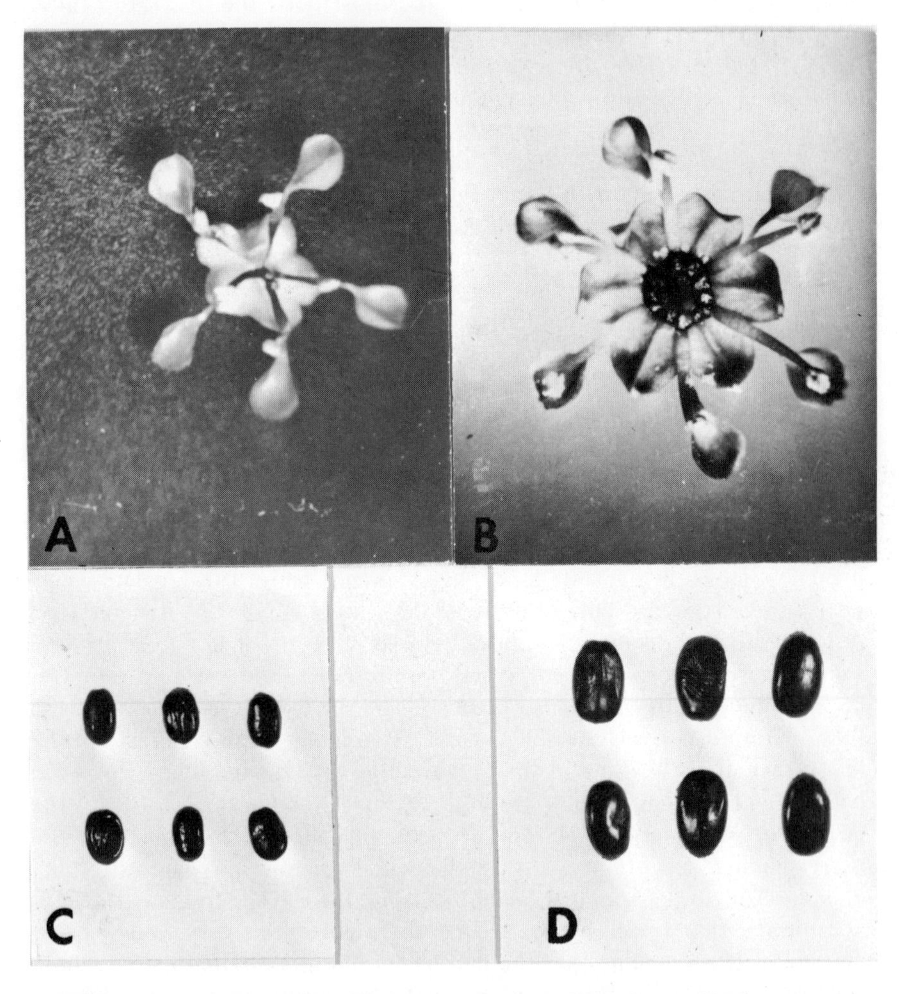

Figure 1. A comparison of flowers and seeds in the genus *Ceanothus*. A. Usual pentamerous flower of the genus *Ceanothus*. B. A six parted flower typical of *C. Jepsonii*. C. Seeds of *C. cuneatus*. D. Larger seeds typical of *C. Jepsonii*.

Ceanothus, the flowers have five sepals, five petals, five stamens, and a three-carpellate ovary containing three seeds. In most individuals of *C. Jepsonii*, however, the number of sepals, petals, and stamens is six or seven, at least in some flowers (Figure 1A, B). Compared to its more common and widespread relative, *C. cuneatus, C. Jepsonii* has considerably larger seeds, containing a larger quantity of stored food material (Figure 1C, D). This would enable the seedling to form a more extensive root system in the earliest stages of its development. While no data have yet been obtained on the root systems of *Ceanothus* species, a generally observed fact is that plants adapted to the drier, unfavorable serpentine soils have extensive root systems. In this example, therefore, natural selection for adaptation to an unfavorable type of soil has probably acted in part through increasing seed size. Since disturbing the architecture of the uniformly three-seeded capsule found in the genus would interfere with the explosive mechanism for seed dispersal which it possesses, the easiest way for selection to produce greater seed size in *Ceanothus* is to increase the size of the capsule as a whole. As in the tomato, some of the genes for increased ovary size apparently exert their effects through increasing the size of the entire floral meristem, and consequently raising the number of floral parts which it can differentiate.

Probable examples of increased numbers of floral parts as adaptively neutral by-products of selection for increase in flower size are found in various plant genera (Stebbins, 1966b). The flowers of the Cactaceae and of the genus *Lewisia* are good examples.

In the examples already mentioned, the side effects of natural selection appear to be adaptively neutral. In other instances, however, some of the genes whose principal effects are adaptive can produce side effects which are harmful to the organism. In such cases, the advantage gained from establishing these genes in the population can be realized only if there are established simultaneously other genes which compensate for harmful side effects. The need for such compensating genes offers particularly favorable situations for evolutionary divergence through selection of different solutions to the same adaptive problem, as postulated by Mayr (1963). An example presented elsewhere is that of compensating genes for capsule size in different flax-inhabiting races of the genus *Camelina* (Stebbins 1950, 1966a).

THE NATURE OF DEVELOPMENTAL SEQUENCES

The importance of gene interaction as a determining factor for natural selection, which is evident in the example just cited, becomes even greater when we become aware of the integrated sequences of gene action which characterize the development of all higher organisms. Three kinds of developmental sequences can be recognized. The first of these is the "biosynthetic pathway," a succession of enzyme-controlled reactions which leads to the formation of a simple organic molecule, such as an amino acid. Biosynthetic pathways have been well known to both biochemists and geneticists for many years, and have been thoroughly explored in both fungi and microorganisms. Their importance to studies of developmental genetics and evolution lies in the fact that two kinds of control mechanisms have been recognized in them. The first is the operator regulator system described by Jacob and Monod (1961; 1963; Jacob, 1966; Monod, 1966). This mechanism includes two genes, the operator and the regulator, whose sole function is to "turn on" and "turn off" simultaneously the entire series of genes which constitute the operon, and which are responsible for the biosynthetic pathway. The second kind of control mechanism, knowñ as feed-back inhibition, turns off the biosynthetic pathway when enough of its end product has been made to satisfy the needs of the organism. It depends upon the fact that the first enzyme in the series possesses a particular sequence of amino acid residues, known as the allosteric site, upon which molecules of the end product can become attached, and which respond to this attachment by turning off the activity of that particular enzyme. The evolutionary significance of these two types of regulation will be discussed below.

Only a small proportion of the integrated gene action which characterizes the development of higher organisms is accounted for by the regulator systems of biosynthetic pathways. A second kind of sequence, which indirectly involves a far greater amont of gene interaction, is based upon interaction between gene products. So far as I am aware, this type of sequence has not been recognized as a class, and has not been given a name. I therefore propose the term "informational relay" for such sequences. An informational relay may be defined as a sequence of interactions between primary, secondary, teritiary, and still more indirect products of gene action, which gives rise to a compound macromolecule or to an organized supramolecular aggregate which functions in cellular metabolism or which directs the organization of cells into tissues.

The shortest, and probably the commonest kind of informational relay, results in the formation of protein molecules, often called dimers or tetramers, which consist of two, four, or more polypeptide chains. Biochemists generally recognize the fact that the gene-coded primary structure of proteins determines not only their secondary and tertiary structure, i.e., their folding pattern, but also the way in which polypeptide chains coded by different genes will become associated to form the quaternary structure of many enzymes and other active protein molecules (Green and Goldberger, 1967). Genetic evidence of this fact is provided by the sickling (S) mutation of hemoglobin. According to Murayama, Olson, and Jennings (1965), the harmful effect of this mutation is that the substitution of valine for glutamic acid at the sixth position of the beta chain causes the alpha and beta chains of the molecule to become associated into an abnormal quaternary structure. This, in turn, greatly reduces the effectiveness of the compound molecule in accepting and releasing oxygen.

Much longer informational relays involve structural proteins, as well as other kinds of molecules, such as lipids and polysaccharides, which are often associated with them in forming cellular organelles and membranes. A generalized informational relay of maximal length is presented in Figure 2. The facts which can be interpreted to support the existence of such relays are numerous, and are derived from observations and experiments on a number of different kinds of cellular structures. The following are some representative examples.

At the level of individual molecules, the most important interrelationships are those responsible for the formation of membranes and organelles. In the case of cell membranes, Green and his co-workers have produced evidence that they consist of repeating units which are held together by hydrophobic attractions inherent in the primary structure of their protein molecules (Green and Goldberger, 1967, p. 223). Each unit, moreover, is believed to contain a specific series of enzymes which fit into

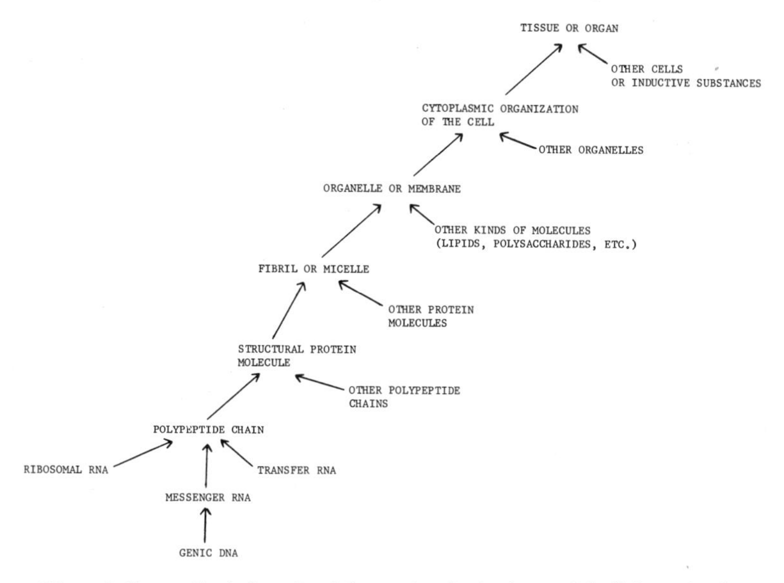

Figure 2. Generalized plan of an informational relay in a multicellular animal.

the matrix of structural protein because of certain specific features of their primary structure. The "fit" between proteins and the lipid molecules of the membrane is less well understood, but may well depend with equal importance upon the structure of the molecules involved. In bacteria, evidence for the importance of molecular interactions in membrane formation is provided by the fact that if cells are deprived completely of their cell wall by lysozyme treatment, they are unable to remake walls even though the necessary cellular materials are synthesized during the growth of the cell (Murray, 1963).

In protozoa, as well as in unicellular algae such as *Chlamydomonas*, there is abundant evidence to indicate that their elaborate organelles are synthesized through the self-assembly of differently constructed proteins which are coded by different genes (Lwoff, 1950; Tartar, 1961; Sonneborn, 1963; Trager, 1965; Nanney, 1966). Some of these genes are certainly nuclear (Lewin, 1954; Sonneborn, 1963), but others may be located in the DNA which is found at the base of the cilia. In these organisms, their entire bodily architecture, including those features responsible for their motility and reaction to stimuli, depends upon the proper interrelationships between different protein molecules, as well as upon cellular conditions which permit their assembly into functional organizations. Among the most important of these cellular conditions are templates formed by previously synthesized molecules.

Similar problems of self-assembly are involved in the synthesis of organelles such as chloroplasts (Bogorad, 1967), and of substances like collagen (Gross, 1961; Gross, Lapiere, and Tanzer, 1963; Weiss, 1962). In the case of collagen, the primary structure of the polypeptide chains appears to be responsible for (a) the linking together of three polypeptide chains to form the helical molecules, (b) the association of molecules to form fibrils, (c) the cross stacking of the fibrils, and (d) the association of apatite crystals with collagen in the formation of bone. The very precise cellular environments required for this orderly succession of processes have been well defined by Gross and his co-workers.

That self-assembly exists in animals at the cell level is evident from a number of experiments involving disassociation and reassociation of cells under controlled conditions (Townes and Holtfreter, 1953; Humphreys, 1963; Hauschka and Königsberg, 1966). While there is at present no evidence which indicates that this self-assembly depends upon the gene-coded primary structure of some of their proteins, this is a likely possibility in view of the known effect of genes upon self-assembly organelles.

The great importance of both a precise cellular environment and a harmonious combination of genes for the normal progress of informational relays is evident from several lines of research. Temperature (Humphreys, 1963), hydrogen ion concentration (Gross, 1961; Townes and Holtfreter, 1953), the presence of certain cations such as calcium and magnesium (Humphreys, 1963; Razin, Morowitz, and Terry, 1965) and various more complex factors of the cellular environment must be precisely balanced with each other. For the assembly of particular kinds of compound enzyme molecules, such as the various isoenzymes of lactic dehydrogenase, the relative concentrations of the various polypeptide chains which compose them are important (Markert, 1963). Environmental factors of the latter sort depend upon differential gene action at earlier stages in development. As a matter of fact, a likely speculation is that the modifying genes well known to geneticists exert their effects largely through modifying the cellular environments which control the behavior of the products of other genes. Evidence on how the course of informational relays can be altered by gene mutation is still scanty. The developmental analysis of mutations which affect chloroplast structure in barley by Wettstein and his associates (Wettstein and Eriksson, 1964) is one of the best examples. They have shown that different mutations can affect development at various stages and in various ways. Chloroplast development must be regarded as a complex succession of events, most of which are directed by remote control. The controlling information is inherent in the primary structure of molecules synthesized by genes located some distance away from the final position of the structural molecules. Integration by remote control is, therefore, a predominant feature of informational relays.

Neither biosynthetic pathways nor informational relays require precisely programmed sequences of gene action which extend over appreciable periods of time. In the case of biosynthetic pathways, all of the enzymes concerned are active at the same time, and the operator-operon relationship ensures the simultaneous activation or repression of the genes which code for all of these enzymes. Informational relays require precise interactions between gene products, but the time when these products are synthesized is immaterial, provided that they are present in the cell in the right concentrations at certain developmental stages. The course of development of any organism, however, of necessity includes such programmed sequences, in which the action of a particular gene follows precisely upon that of another, otherwise unrelated gene. That such sequences of gene action depend largely upon differential activation and repression of specific genes is now evident (Beermann, 1963). We must consider, therefore, a third type of developmental sequence, termed an "epigenetic sequence." How such sequences are integrated is, however, still a matter of speculation. Several different mechanisms may be invoked. That the action of hormones is involved in some epigenetic sequences is reasonably clear (Beermann and Clever, 1964). Moreover, the conclusion is almost inescapable that the end products of certain biosynthetic pathways can either repress the activity of the enzymes involved in other biosynthetic pathways responsible for earlier developmental stages, or that they can activate subsequent pathways. Grobstein (1964) has suggested that embryonic induction can be effected by substances which act on the nuclei of cells in the responding tissue. Koch and King (1964) have studied a gene in *Drosophila melanogaster* which apparently produces a substance that causes cells to stop dividing and begin differentiating.

Two rather similar models have been produced for the activation and regulation of epigenetic sequences. Halvorson (1965), based upon his studies of *Bacillus subtilis*, has suggested that the sequence responsible for sporulation in this species consists of a series of individual biosynthetic pathways which are linked together by the action of inhibitor and activator substances produced by each pathway. Stern (1963), in order to explain the carefully programmed epigenetic sequence taking place in the premeiotic and meiotic cells of a higher plant (*Trillium*), has postulated a similar mechanism, except that the sequential units are not individual pathways but complexes of operons which he designates "macrooperons." Clearly, the nature of epigenetic sequences is one of the most important unsolved problems of biology, and one which is of vital importance to students of evolution.

DEVELOPMENTAL INTEGRATION AND INTERNAL SELECTION

One might ask at this point, what bearing do all these facts have upon the relationship between natural selection and evolutionary progress? The answer to this question is that the more we recognize the importance of coordinated, integrated gene action in development, the more we realize that the internal, genotypic environment provided by the organism can exert an equal or greater selective pressure upon newly appearing gene mutations than does the external environment. Moreover, this pressure can be exerted soon after the appearance of the mutation. Dominant mutations which affect the integration of cellular metabolism will be cell lethal, while recessive mutations of this kind, appearing first as homozygotes in a zygote, will produce death of embryos in the earliest stages of development. In either case, the loss to the evolving line is minimal. Compensation for losses due to the lethal effects of dominant mutations can be made by regenerative growth, and for the effects of recessive mutations by the extra reproductive potential which every species possesses.

Recent comparative studies of enzyme molecules give us a chance to estimate a minimum proportion of possible mutations which are rejected by internal selection. The rationale of such estimates is as follows. As Mayr (1965) has pointed out, it is unreasonable to expect that a particular triplet sequence of DNA will be completely incapable of mutating during periods of time which extend over millions of years. Hence, any sequence of amino acids possessed in common by two organisms whose common ancestor existed millions of years ago must owe its constancy to the fact that all mutations which might alter this sequence have been uniformly rejected by natural selection.

Using this criterion, we can make significant comparisons between such molecules as cytochrome *c*, about which much information has been accumulated, and is summarized by Margoliash and Smith (1965) as well as by Jukes (1966). Human cytochrome *c* contains 104 amino acid residues. When this molecule is compared with cytochrome *c* of the horse, the number of residues is the same in both molecules. Moreover, 92 out of the 104 consist of identical residues in the same positions. This means that except for variations due to the degeneracy of the code, 276 out of the 312 base pairs must be the same in the two molecules. By comparing the 12 triplet codons which differ in the two molecules, Jukes estimated that the differences between human and horse cytochrome *c* could be explained by mutations at only 18 of the possible 312 sites. If we assume that mutations involving single base pairs can occur with equal frequency at any site, then we must conclude that over 90 per cent of the mutations which have occurred in the gene coding for cytochrome *c* throughout the course of evolution of horse and man from their common ancestor have been rejected by selection. Moreover, this is only a minimal estimate, since even at the sites which have undergone mutation, only some of the possible mutations have occurred and become established.

Whether or not this figure of 90 per cent rejection of mutations during the evolution of mammalian orders holds for most genes and the molecules for which they code will not be known until many more data are available. Nevertheless, any divergent figures which may emerge from studies of other molecules will not affect the fact that the number of mutational sites which are constant over millions of generations far exceeds the number of those which are concerned directly with the chemical reaction catalyzed by the enzyme, i.e., with its active and binding sites. Hence, many regions of the gene cannot mutate successfully for reasons other than the interference of such mutations with the catalytic activity of the enzyme. The constancy of such regions is probably due for the most part to their essential role in integrating the molecule concerned with the other molecules of the cell. A specific example of such integration is suggested by Margoliash and Smith (1965). They point out that all known molecules of cytochrome *c*, which are derived from organisms as different from each other as yeast, tuna fish, and man, possess exactly the same sequence of amino acid residues at positions 70 to 80, which are far from the active sites of the molecule. They suggest that this invariant sequence may bear a specific relationship to a coenzyme which joins with cytochrome *c* in all of its reactions, or that it may determine the placement of the molecule upon the mitochondrial membrane where it does its work.

Although the demands of developmental integration at the cell level greatly reduce the number of successful mutations which are possible, they cannot reduce this number to such an extent that this factor alone will guide evolution in a particular direction. For instance, if there remain only 100 sites at which successful mutations can occur, and four different bases can exist at each site, then the number of possible gene combinations based upon these differences is 4^{100}, a number far exceeding that of every living organism upon the earth. There is, therefore, plenty of leeway for the more conventional action of the external environment as a selective agent.

SELECTIVE INERTIA AS A DIRECTIVE FORCE

In multicellular animals and plants, the need for internal harmony at the level of the organ and the integration of the organism as a whole places restrictions upon the success of mutations and gene combinations which are even more significant than are those which operate at the chromosomal or cell level. For example, the diversity of evolutionary pathways which insects have taken is far greater than in any other group of organisms. Nevertheless, no group of insects was able to evolve into forms capable of occupying the niche now occupied by passerine birds. This niche was completely open and unoccupied when the orders of insects were evolving. One reason for this failure is that if insects were to become as large and swift-flying as a sparrow, their external skelton would not be strong enough to support the wing muscles for flight, and at the same time be light enough for the necessary byoyancy. Only the internal bony skelton of vertebrates makes this possible. Because of their external skelton, as well as the nature of their circulatory and nervous system, the evolution of insects has been canalized into small size. In all probability, moreover, the evolution of individual groups of insects is even more restricted by canalization. For instance, among the more than 1,000 species of the genus *Drosophila*, which are world-wide and occupy an enormous range of habitats, none has evolved to a size as large as that of a house fly, or as small as a gnat. The reasons for this are probably to be sought in the effect upon mutation of the demands for bodily harmony at the level of the tissue or organ, which could probably be discovered by careful anatomical, physiological, and developmental studies of the organisms concerned. From these examples, we must conclude that the existing organization of any population of organisms, which has been determined by previous actions of natural selection, greatly restricts the number of evolutionary pathways which lines descended from them can follow.

This restriction is augmented by the fact that given a particular pattern of adaptive organization, and a new environmental factor to which several kinds of adaptive responses might be possible, the new adaptation which is actually acquired will be that which requires the smallest mutational readjustment of the existing developmental pattern. In other words, natural selection for a general property, such as increased fecundity, will follow the lines of least resistance. I have illustrated this point elsewhere by referring to the different ways by which natural selection for increased seed production could affect the architecture of a flowering plant (Stebbins, 1966b). An increased number of seeds per plant could be acquired by an increase in either (1) the number of seeds per carpel, (2) the number of carpels per flower, (3) the number of flowers per inflorescence, or (4) the number of inflorescences or branches per plant. In a plant having a small number of large flowers and few carpels per flower, such as a lily or tulip, the path of least resistance is via an increase in number of seeds per carpel. If the flower is large and already has several one-seeded carpels, as in a magnolia, increase in number of carpels per flower is the easiest response. Finally, if, as in the sunflower family (Compositae), each flower has a precisely programmed development leading to only one seed per flower, increased seed production is acquired most easily by increasing the number of flowers per inflorescence or head, or the number of heads per plant.

This tendency for selection to follow the lines of least resistance I propose to call the principle of "selective inertia." Its relationship to evolutionary canalization may be expressed as follows. In any evolving line represented by related populations of organisms, the future direction of evolution is restricted and canalized by virtue of the particular adaptive complex of characters possessed by the population at any particular time. This canalization usually becomes greater as lines evolve, and often leads to the extinction which follows inevitably upon overly narrow specialization. Only occasionally does a canalized line overcome its selective inertia, break out of its canalization, and produce derived lines which radiate in new adaptive directions.

In relation to trends toward greater evolutionary complexity, the most significant fact about both developmental integration at the molecular level and selective inertia at higher levels is that they usually discriminate against simplification or the elimination of developmental stages. By doing so, they produce a bias in favor of increasing complexity. Mutations which destroy molecular integration are much more likely to be cell-lethal, and so to disappear quickly and with little damage to the organism, than are those which might lead to more complex integrated molecular interactions in the future. At the chromosomal level, deletions which remove or inactivate the genes coding for any vital enzyme of a sequence are almost certain to be cell-lethal, while duplications of such genes cannot only be tolerated but also may in many instances contribute significantly to evolution toward greater developmental complexity. In fact, the evolutionary contribution of such duplications deserves special attention.

THE EVOLUTIONARY ROLE OF CHROMOSOMAL DUPLICATIONS

The hypothesis that specific gene duplications have promoted the evolution of organisms toward greater developmental and organizational complexity dates back to the early days of research on *Drosophila*, having been put forward in 1918 by Calvin Bridges and reaffirmed in 1935, when the banding pattern of salivary chromosomes became known (Bridges, 1935). The same opinion has been expressed by Weir (1946), Lewis (1951) , and others. The argument runs as follows. The gene molecule is so complex that many different variants of its structure, or isomers, are possible. Most of these function poorly or not at all. Hence, the transition by mutation from one functional isomer of a gene to another is likely to be via one or more mutations which destroy or greatly weaken its function. In the case of genes having key functions for cell metabolism, these inadaptive "valleys" can be traversed most easily if two genes having the same function are present, so that while one of them is being reorganized, the other can carry on its essential function.

The discoveries of molecular genetics have greatly strengthened this hypothesis. A gene possesses hundreds of mutational sites, and the number of possible isomers, or sequential orders of its nucleotides, is almost infinite. Of these, only a small proportion can code for a functional enzyme or polypeptide chain. At the level of the enzyme molecule, we now recognize that many, and perhaps most enzymes consist of two or more different kinds of polypeptide chains, coded by as many different genes, and that the primary structure of these chains is so similar that the genes which code for them most probably are descended from a single ancestral gene. Furthermore, higher animals possess active proteins with similar functions, such as hemoglobin and myoglobin, which are apparently also coded by homologous genes (Ingram, 1963). The same is true of various families of isoenzymes, for which in at least some instances, homologous genes which reflect past duplications may be present.

How great a role have such tandem duplications played in the evolution of organisms toward greater complexity? One criterion of its magnitude is the extent to which evolution toward greater organizational and developmental complexity has been correlated with increase in nuclear DNA content, which measures at least approximately the amount of gene duplication. This correlation can be roughly estimated from published summaries of the DNA content in the nuclei of various organisms, such as those of Sinsheimer (1957), Sparrow and Evans (1961), and myself (Stebbins, 1966c). From these we can conclude that the evolutionary progression from bacteria to primitive eucaryotes, such as flagellate protozoa, primitive algae, and sponges, was accompanied by a five- to ten-fold increase in the amount of DNA per nucleus. From these primitive eucaryotes to the most complex of modern organisms, such as land vertebrates and seed plants having low chromosome numbers and moderate-sized chromosomes, the increase in DNA content per nucleus was about the same order of magnitude. This latter increase was, however, by no means regular. Among the vertebrates, by far the highest contents of DNA are possessed by the nuclei of such relatively primitive forms as lungfishes and amphibia. Among vascular plants, tenfold differences exist between different members of the same or related classes, orders, or even families, such as grasses and legumes.

What general conclusions can we reach from these data? The first must be a caution: Before any firm conclusions can be made, many more data are needed on the nuclear DNA content of different organisms. Nevertheless, the considerable increase which has taken place during the progression from bacteria to higher vertebrates and seed plants suggests that duplication and subsequent differentiation of genes has played an important role in this progress. We can be equally sure that many instances of greatly increased DNA content, such as those found in amphibia, have not been accompanied by progression toward increased complexity. These latter increases may have possessed adaptive advantages for other reasons, which at present are completely unknown. I have made some speculations about them elsewhere (Stebbins, 1966c).

The probability that the opportunity for evolving greater complexity is by no means the only advantage inherent in an increased nuclear DNA content should be regarded as favorable rather than inimical to the hypothesis that this increase can promote evolutionary progress. The advantages of such progress can be realized only long after the duplication has occurred. Hence, unless there is a different, more immediate advantage to a duplication at the time of its occurrence, it would have difficulty in spreading and becoming established in a series of populations.

Although the gene duplication-differentiation cycle may well account for evolutionary progress from primitive bacterial ancestors to the first representatives of the more advanced phyla and classes, it cannot account for evolutionary advancement within these modern groups. Among their orders and families, there is no correlation whatever, either positive or negative, between degree of organizational or developmental complexity and amount of DNA per nucleus. How can we account for this apparent paradox? One suggestion is as follows. The progressive evolution at the level of the origin of new phyla, which is in general correlated with increase of DNA content, is accompanied by the appearance of many new kinds of cells, tissues, and organs. This increase in the numbers and kinds of differentiated structures probably requires also an increase in the number of different enzyme systems, or at least of different isoenzymes capable of performing similar functions in a variety of different cellular environments. On the other hand, evolution within the more advanced phyla, such as chordates and seed plants, consists largely of changes in the amounts of the different kinds of tissues, as well as in the sizes and shapes of different organs, and the relationships between them. Consequently, we might conclude that progressive evolution within these higher groups has consisted not so much of the addition of metabolic functions or of processes of differentiation, but of greater, more complex integrations of these functions, and of alterations in the sizes and conformations of various organs. Such alterations could be effected by mutations and recombinations of genes, already present, which have specific regulating and coordinating functions.

As an illustration of this latter kind of change, we can take the example of the great increase in brain size which took place in the hominid line between the australopithecine stage and that of *Homo sapiens*. This increase carried with it a complete change in the adaptive properties of the organisms concerned. Yet at the level of the cells, tissues, and organ, the change was minimal. It could have been accomplished by a shift in regulatory mechanisms active during early embryo development, which caused various cells of the head region to proliferate more rapidly than other cells of the body. The profound difference between this kind of evolutionary change and that which, early in the evolution of life, brought into being the various cellular organelles with their attached enzyme systems, should be obvious to any biologist. Clearly, moreover, all of the evolutionary changes which have brought into being the various orders of mammals were much more like the alterations of the head in hominid evolution than the cytological and histological changes which accompanied the evolution of the major phyla.

CONCLUSION

The Basis of Increasing Complexity

In conclusion, I should like to give a tentative answer to a question which evolutionists have asked ever since Darwin introduced the concept of natural selection as a guiding force, and which was recently emphasized by Medawar (1966). This is, if the direction of evolution is determined by more or less random conditions and changes in the external environment, acting upon a random lot of genetic mutations and recombinations, how do we account for the apparent trend of evolution toward greater complexity?

The first part of my answer is that this trend is only apparent, if by the word trend we mean the action of some force which inevitably, sooner or later, compels living organisms to evolve into more complex forms. That such a compelling force does not exist is apparent from the fact that organisms like bacteria and blue-green algae have perpetuated their evolutionary lines for at least three billion years without evolving into anything more complex. Moreover, even in those phyla which we regard as progressive, the proportion of species which have evolved into more complex adaptational types is minuscule compared to those which either became extinct or evolved into forms not very different from themselves. This is evident from the fact that most of the more advanced groups are monophyletic in origin, although the simpler groups from which they came must have contained many different species, and many relatively simple species belonging to these groups have persisted to the present time. Since they first appeared on the earth, the group of flagellate protista (Protozoa and Algae) must have evolved hundreds of thousands of different species, but not more than fifty to a hundred of these were ancestral to more highly organized groups. Among the hundreds of species of jawless vertebrates which dominated the seas during the Ordovician and Silurian periods, only one or two of them gave rise to jawed fishes. Reptiles by the tens of thousands dominated the earth during the Mesozoic era, but only two lines, one belonging to a small group of the therapsid order and the other to a small, inconspicuous group related to the dinosaurs, became the ancestors of more complex, highly organized classes of vertebrates, the mammals and birds. Looking at the whole sweep of evolution, one sees not progressive trends, but rather successive cycles of adaptive radiation followed by wholesale extinction, or by widespread evolutionary stability. The ascent to new levels of complexity has been accomplished by a few exceptional groups. As Simpson (1953) has so clearly pointed out, evolution has consisted not of a series of inevitable trends, but has been highly opportunistic. Nevertheless, although the ascents to greater complexity have been rare, the descents to greater simplicity have been even rarer so that the net over-all result has been increase in complexity. This is the fact which we need to explain.

The explanation which I would like to suggest is a summary of the argument presented in my previous remarks. While the effects of mutations are at random relative to the adaptiveness of the organism, those mutations which survive to the extent that they can become established even temporarily in populations are not at random relative to the genetic and developmental integration of the organism in which they occur. Those mutations which tend to destroy this integration are quickly eliminated, and most often do not even reach the point where they affect the interaction between the adult individual and its environment. Those which either do not affect this integration, or which make possible the evolution of still more complex integrations, are much more likely to become established in populations than are those with destructive or inactivating effects. Some of these new genes, in combination with each other, will, from time to time, lift an evolutionary line up to a new level of complexity in adaptation. In the long run, those complexities which have been most successful are the ones which have rendered organisms increasingly independent of the immediate, short-term fluctuations of their environment. They have enabled such progressive lines to evolve in directions which have been determined to an increasing degree by the kind of organized structure which they themselves have previously evolved. In this way, direction, continuity, and progression have become part of the evolutionary process.

LITERATURE CITED

Beerman, W., 1963, "Cytological Aspects of Information Transfer in Cellular Differentiation," *Amer. Zool.*, 3:23–32.

Beerman, W., and U. Clever, 1964, "Chromosome puffs," *Sci. Amer.*, 210:50–58.

Bogorad, L., 1967, "The Organization and Development of Chloroplasts," p. 134–185. In

John M. Allen, ed., *Molecular Organization and Biological Function*, Harper and Row, New York.

Bridges, C. B., 1935, "Salivary Chromosome Maps," *J. Hered., 26:60–64.*

Green, D. E., and R. F. Goldberger, 1967, *Molecular Insights Into the Living Process*, Academic Press, New York.

Grobstein, C., 1964, "Cytodifferentiation and Its Controls," *Science*, 143:643–649.

Gross, J., 1961, "Collagen," *Sci. Amer.*, 204(5):120.

Gross, J., C. M. Lapiere, and M. L. Tanzer, 1963, "Organization and Disorganization of Extracellular Substances: The Collagen System," p. 175–202. In M. Locke, ed., *Cytodifferentiation and Macromolecular Synthesis*, Academic Press, New York.

Halvorson, H. O., 1965, "Sequential Expression of Biochemical Events During Intracellular Differentiation," Symp. Soc. General Microbiol. XV. *Function and Structure in Microorganisms*, p. 343–368.

Hauschka, Stephen D., and Irwin R. Königsberg, 1966, "The Influence of Collagen in the Development of Muscle Clones," *Proc. Nat. Acad. Sci.*, 55:119–126.

Houghtaling, H., 1935, A Developmental Analysis of Size and Shape in Tomato Fruits," *Bull. Torrey Bot. Club*, 62:243–252.

Huether, C. A., 1966, "The Extent of Variability for a Canalized Character (Corolla Lobe Number) in Natural Populations of *Linanthus* (Benth)," Ph.D. Thesis, University of California, Davis.

Humphreys, T., 1963, Chemical Dissolution and *in vitro* Reconstruction of Sponge Cell Adhesions; Isolation and Functional Demonstration of the Components Involved," *Dev. Biol.*, 8:27–47.

Ingram, V. I., 1963, *The Hemoglobins in Genetics and Evolution*, Columbia University Press, New York.

Jacob, Francois, 1966, "Genetics of the Bacterial Cell," *Science*, 152:1470–1478.

Jacob, F., and J. Monod, 1961, "Genetic Regulatory Mechanisms in the Synthesis of Proteins," *J. Molec. Biol.*, 3:318–356.

Jacob, F. and J. Monod, 1963, "Genetic Repression, Allosteric Inhibition, and Cellular Differentiation p. 30–64. I M. Locke. ed., *Cytodifferentiation and Macromolecular Synthesis*, Academic Press, New York.

Jukes, T. H., 1966, *Molecular and Evolution*, Columbia University Press, New York.

Koch, E. A., and R. C. King, 1964, "Studies on the FES Mutant of *Drosophila melanogaster*," *Growth*, 28:325–369.

Lerner, I. M., 1954, *Genetic Homestasis*, John Wiley and Sons, New York.

Lewin, R. A., 1954, "Mutants of *Chlamydomonas moewusii* with Impaired Motility," *J. Gen. Microbiol.*, 11:358–363.

Lewis, E. B., 1951, "Pseudoallelism and Gene Evolution," Cold Spring Harbor Symp. Quant, Biol., 16:159–174.

Lwoff, A., 1950, *Problems of Morphogenesis in Ciliates: The Kinetosomes in Development, Reproduction, and Evolution*, John Wiley and Sons, New York.

Margoliash, E., and Emil L. Smith, 1965, "Structural and Functional Aspects of Cytochrome *c* in Relation to Evolution," p. 221–242, In V. Bryson and H. G. Vogel, eds., *Evolving Genes and Proteins*, Academic Press, New York.

Markert, C. L., 1963, "Epigenetic Control of Specific Protein Synthesis in Differentiating Cells, p. 65–84, in M. Locke, ed., *Cytodifferentiation and Macromolecular Synthesis*, Academic Press, New York.

Mayr, E., 1963, *Animal species and Evolution*. Harvard University Press, Cambridge.

———, 1965, Discussion of Part IV, p. 293–294, In V. Bryson and H. J. Vogel, eds., in *Evolving Genes and Proteins*, Academic Press, New York.

Medawar, Sir Peter, F.R.S., 1966, "A Biological Retrospect," *BioScience*, 16:93–96.

Monod, J., 1966, "From Enzymatic Adaptation to Allosteric Transitions," *Science*, 154: 475–483.

Murayama, M., R. A. Olson, and W. H. Jennings, 1965, "Molecular Orientation in Horse Hemoglobin Crystals and Sickled Erythrocytes," *Biochim. Biophys. Acta*, 94:194–199.

Murray, R. G. E., 1963, "The Organelles of Bacteria," p. 28–52, in D. Mazia and A. Tyler, eds., *General Physiology of Cell Specialization*, McGraw-Hill, New York.

Nanney, D. L., 1966, "Corticotype Transmission in *Tetrahymena*," *Genetics*, 54:955–968.

Razin, Shmuel, Harold J. Morowitz, and Thomas M. Terry, 1965, "Membrane Subunits of *Mycoplasma laidlawii* and Their Assembly to Membrane-like Structures," *P.N.A.S.*, 54:219–225.

Simpson, G. G., 1953, *The Major Features of Evolution*, Columbia University Press, New York.

Sinsheimer, R. L., 1957, "First Steps Toward a Genetic Chemistry," *Science*, 125:1123–1128.

Sonneborn, T., 1963, "Does Preformed Cell Structure Play an Essential Role in Cell Heredity," p. 165–221, in J. M. Allen, ed., *The Nature of Biological Diversity*, McGraw-Hill, New York.

Sparrow, A. H., and H. J. Evans, 1961, "Nuclear Factors Affecting Radiosensitivity. I. The Influence of Nuclear Size and Structure, Chromosome Complement, and DNA Content," Brookhaven Symp. Biol., 14:76–100.

Stebbins, G. L., 1950, *Variation and Evolution in Plants*, Columbia University Press, New York.

______, 1966a. *Processes of Organic Evolution*, Prentice-Hall, Englewood Cliffs, N.J.

______, 1966b, "Adaptive Radiation and Trends of Evolution in Higher Plants," p. 101–142, in Th. Dobzhansky, M. K. Hecht, and Wm. C. Steere, eds., *Evolutionary Biology*, Appleton, Century, Crofts, New York.

______, 1966c, "Chromosome Variation and Evolution," *Science*, 152:1463–1469.

Stern, Herbert, 1963, "Intracellular Regulatory Mechanisms in Chromosome Replication and Segregation," *Fed. Proc.*, 22:1097–1102.

Tartar, V., 1961, *The Biology of Stentor*, Pergamon Press, New York.

Townes, P. L., and J. Holtfreter, 1953, "Directed Movements and Selective Adhesion of Embryonic Amphibian Cells," *J. Exp. Zool.*, 128:53–120.

Trager, William, 1965, "The Kinetoplast and Differentiation in Certain Parasitic Protozoa," *Amer. Natur.*, 99:255–266.

Waddington, C. H., 1962, *New Patterns in Genetics and Development*, Columbia University Press, New York.

Weir, J. A., 1946, "Sparing Genes for Further Evolution," Proc. Iowa Acad. Sci., 53:313–319.

Weiss, P., 1962, "From Cell to Molecule," p. 1–72, In John M. Allen, ed., *The Molecular Control of Cellular Activity*, McGraw-Hill, New York.

Wettstein, Diter von, and Gosta Eriksson, 1964, "The Genetics of Chloroplasts," p. 591–612, in *Genetics Today*, Proceedings of the XIth International Congress of Genetics, Pergamon Press, Oxford.

-3-

The Paradigm for the Evolutionary Process

C. H. WADDINGTON
Institute of Animal Genetics
Edinburgh, Scotland

It is essential that biologists should be as careful and precise as possible in formulating an adequate and profound theory of evolution, since it is now widely recognized that this is the most central theory in the whole of biology. In the first quarter of this century the most influential biologists considered that the basic characteristic of life is to be sought in the metabolic activities of living things. A biological system, they pointed out, is one which takes in relatively simple substances from its surroundings and elaborates these into more complex substances. It appears to operate in a manner directly opposed to the second law of thermodynamics. Later it was realized that these apparent contradictions of fundamental physical law are only local and temporary, and, indeed, occur only in parts of the total system and not in the complete system which comprises both the living organism and its environment. The search for a fundamental theory of biology then shifted towards genetics. A living system began to be regarded as basically one in which there is hereditary transmission of mutable information. This makes possible, or indeed inevitable, the process of natural selection, and thus of an evolutionary process which can be held responsible for all the further elaboration which we find in the living world today.

This Neo-Darwinist view of the basic nature of life is the dominant one at the present time. It is the view most generally discussed by the people interested in scientific theory, most of whom approach the subject primarily from the side of physics. In my opinion, however, it is inadequate in several ways, and I should like to discuss some of these.

THE NECESSITY OF THE PHENOTYPE

The term "Neo-Darwinism" is often applied very loosely, to include almost any recent evolutionary theorizing which has been influenced by Mendelian genetics. Many of the successes claimed for it by its defenders (e.g., the understanding of evolution in social insects) should, in fact, be attributed simply to Mendelism itself rather than to Neo-Darwinism. In its strict sense, it refers to the mathematical models developed originally by Sewall Wright, Haldane, and Fisher, and elaborated since then by many later authors. This strict Neo-Darwinism does not involve any necessity to refer to the phenotype. It postulates a population whose individuals possess genotypes selected out of a collection of genetic units (genes, cistrons) G_1, G_2, . . . G_n. These are to be replicated by some process, which produces a new genotype consisting of units G'_1, G'_2, . . . G'_n . It is supposed that if one unit, G_j , undergoes a change (mutation) into a new form $G_{j'}$, this alteration will be copied by the replication process so as to produce a new unit $G_{j'}$. In order to bring about a process of evolution, it is necessary to make one further assumption, namely that there are interactions with the environment which bring about differences in the frequencies with which G_j and $G_{j'}$ are represented in later generations. These interactions constitute the process of natural selection. Provided that this occurs, and that there is transmission of mutable hereditary information, then evolutionary changes, in the sense of alterations in the frequencies of G_j and $G_{j'}$ in the population, are inevitable.

This statement makes no reference to anything which needs to be considered a phenotype as opposed to the genotype. Neo-Darwinist evolution is, therefore, theoretically possible in systems in which no such distinction is called for.

Now, in practice we do not actually find any such systems in the biologically evolving world, which relies mainly, if not exclusively, on a carrier of hereditary information, DNA, which can be replicated only with the aid of a phenotype consisting of enzymes, such as polymerases, and those in the biosynthetic pathways producing necessary subunits (nucleotides, sugars, ATP, etc.). There are indeed systems which can be thought of as exhibiting hereditary transmission of mutable information without the participation of a phenotype, but they are normally classified as belonging to the inorganic world. An example, discussed by Cairns Smith (1965) is the transmission of imperfections of the lattice during the growth of a crystal. He has suggested that a process akin to natural selection would occur in such situations, and that there might be an effective "pre-biotic evolution." It is, however, by no means clear how this "natural selection" could amount to anything more than differences in the reliability with which the imperfection is transmitted; and this would be dependent on the nature of the imperfection itself, not necessarily involving any interaction between the hereditary information and the environment. Such systems do not, for this reason, provide true models of biological evolution for which it is essential that it be the environment (i.e., something outside the organism) which exerts natural selection by determining the frequencies with which alternative hereditary units are represented in later generations.

Systems in Neo-Darwinist evolution must find some way of reconciling two rather conflicting requirements: (a) they must have a method of storing genetic information in a form which is sufficiently unresponsive to environmental influences to be reliable, and (b) they must interact with the environment sufficiently to feel the effects of environmentally-directed natural selection.

Any system which incorporated both these requirements into a single substance, which acted both as memory-store and as environment-reacter, would almost certainly have to exhibit Larmarckian effects in which the environment could directly produce changes in the content of the stored genetic information. It is, perhaps, not necessary to discuss further how far it might be theoretically possible to design such a system, since we know that this is not the method that the living world has adopted. It has followed the strategy of storing genetic information in a chemically unreactive substance, DNA, which has minimal interactions with the environment, and developing from this (through the intermediate step of RNA) a phenotype based upon the highly reactive proteins, which both exert influences on and are influenced by the environment.

Even if ways can be found around the conclusion that the existence of phenotypes in systems undergoing evolution is a logical necessity, it would remain true that it is such a general characteristic of the evolutionary process actually employed by the biological world that no theory which omits it can be considered adequate. The model of evolution described in the first paragraph of this section, which was used for the formulation of the mathematical treatments of Wright, Haldane, and Fisher, suffers from this deficiency.

The introduction of phenotypes into this system of ideas involves a rather radical re-casting. It is sometimes considered enough merely to insert into the old equations a "fudge factor," the "heritability," to allow for the fact that the phenotypes on which natural selection acts give only partial information about the genotypes. But this is not sufficient to encompass the necessary complexity of the situation. A phenotype is reactive to the environment in several different ways. It is not only responsive to the fundamental process of natural selection, (i.e., to influences on the rate at which it transmits its genotype to later generations), but it also submits to actions which modify the epigenetic processes by which the character of the phenotype becomes determined during development; and it may also act positively on the environment so as to alter it.

The last of these, the influence of the phenotype on the environment, can perhaps be neglected in a treatment of general principles, since it is of importance mainly in the later stages of evolution, when organisms have attained some complexity. But the modifications of the character of phenotypes by environmental effects cannot be left out of account, since perhaps the major problem of the whole of evolutionary theory is to account for the adaptation of phenotypes to environments.

THE NECESSITY TO CONSIDER MORE THAN ONE ENVIRONMENT

The strict Neo-Darwinist paradigm is unsatisfactory in another respect, namely, that it involves only one uniform environment, through which natural selection is exerted in a form which requires specification by one single coefficient for each type of biological entity. Again, as with the omission of the phenotype, there are several different objections to this—though perhaps they should be regarded as different aspects of a basic general objection.

To put the matter abstractly first: There are only two sources of evolutionary change; alterations in the environment, or alterations in genes. A paradigm in terms of a single uniform environment implies either attainment of an equilibrium, or evolutionary changes brought about by the appearance of new genes. But the latter is a very weak prop to rely on, since it is normally held that all possible mutations are constantly occurring at definite frequencies. One could perhaps escape this dilemma by appealing to rare mutational events involving large-scale restructuring of the genotype (additions, deletions, inversions, etc.), or rare incorporation of large masses of genetic information by processes such as introgressive hybridization, incorporation of episomes, etc., but this would be an uncomfortable basis for a general theory of evolution.

On a more pragmatic level, one may ask whether the concept of a single uniform environment is ever even conceivably applicable to the real world, which appears inescapably heterogeneous. And if it is not, it should be remembered that evolution provides mechanisms by which any initial inhomogeneity will become either exaggerated in kind or increased in the number of sub-regimes. For instance, if we start with a total universe containing two environmental regimes (niches) A and B, each dominated by a biological species A′ or B′, then it will always be possible for some evolutionary descendant of one or other of these species to delimit as its own niche some appropriate function of the previously existing entities, F(A,B,A′,B′); indeed, there is an infinite set of such functions to be used in this way. This is the general explanation for one of the features of evolution which seems to prove most puzzling to physical scientists, who ask why such an enormous variety of different types should have been produced, although the existence of primitive organisms such as bacteria, at the present day, proves that they are functionally quite "fit" enough to survive. The point is that their mere presence opens the possibility that something will evolve to exploit them, e.g., as a consumer of them. The utilization, as an environmental niche, of a function of pre-existing species and niches F (A,B,A′,B′) will not always demand a biological organization more complex than that of A′ and B′; a mere parasite on A′ may be much less complex or highly evolved than its host. But in general terms one may expect that, among the whole array of functions F which are potentially utilizable, some will actually be used which do require the elaboration of more complex reaction and control systems than any involved in the entities of which the function is composed. We would thus expect to find, not that the evolution of living systems exhibits any universal program of "progressive change" (in the sense of increase in complexity or the like), but that there would be a tendency for the highest degree of complexity reached to increase gradually as the process continues. If the primitive biological world was populated only with bacteria, say, it is, for the reasons just given, difficult to imagine it remaining with no more complex organisms appearing within it, even if the combinatorial possibilities of the genes in the bacteria were sufficient to keep a process of evolutionary change on the move at the bacterial level from those times to the present.

BASIC EQUATIONS OF THE NEO-DARWINIAN AND POST-NEO-DARWINIAN PARADIGMS

The paradigm of the fundamental evolutionary process which has been put forward above leads to an algebraic formulation rather different than that of the Neo-Darwinian scheme, and there are differences in the types of question which can be meaningfully posed in the two types of situation. The Neo-Darwinian algebra was most clearly and simply stated by its originator, J. B. S. Haldane (1924), and for purposes of contrasting the paradigms we may imitate the extreme simplifications of his exposition.

Clonal Reproduction

The simplest possible evolving system will comprise two vegetatively reproducing haploid clones, with genotypes A and a. In the Neo-Darwinian scheme, these exist in one environment S. The action of natural selection is represented by assigning to one clone, say A, a selection coefficient of 1, while the other clone a has a coefficient of $1 - k$. Then clone A will increase in frequency in the mixed population, and clone a will decrease in frequency.

In the "post-Neo-Darwinian" scheme, we have to consider not only two clones A and a, but also two environments X and Y, which can affect the phenotypes of members of A and a developing within them. Moreover, we have to allow for the possibility that an organism developing mainly in one environment may be selected mainly in the other. For simplicity, we may at this stage separate these two actions of the environment, and consider the organisms as classifiable into those developed in X and selected in X, those developed in X and selected in Y, and so on. Thus, for each clone there are four classes to which selection coefficients have to be assigned. The behavior of the system will depend on the attribution of the coefficients to the classes. The kind of questions we can ask of the algebra concern the evolutionary advantages and disadvantages of various principles which might govern the magnitude and distribution of the coefficients.

Suppose there are two clones A and a, and two environments X and Y, with frequencies p and $1 - p$. As case 1 let us assume that a proportion q of organisms is selected in X and $1 - q$ in Y. Then for each clone we have:

	Developed in X Selected in X	Developed in X Selected in Y	Developed in Y Selected in X	Developed in Y Selected in Y
frequency	pq	$p(1 - q)$	$q(1 - p)$	$(1 - p)(1 - q)$

Suppose clone a has perfect adaptiveness, i.e., always has full natural selective efficiency in the environment in which it developed. Then its coefficients would be:

$$1 \qquad 1 - k_1 \qquad 1 - k_2 \qquad 1$$

But let clone a be fully canalized for X, i.e., show full natural selective efficiency in X whatever environment it had developed in. Its coefficients would be:

$$1 \qquad 1 - k_3 \qquad 1 \qquad 1 - k_4$$

Simplifying further, we may assume that the chance of being selected within an environment is proportional to the frequency of that environment, i.e., $q = p$. Further, let us take $k_1 = k_2 = k$ and $k_3 = k_4 = k'$.

Then if the frequencies of A and a in generation n were $1 - u$ and u, respectively, in generation $n + 1$ they will be:

$$A_{n+1} = (1 - u)[p^2 + (1 - p)^2 + 2p(1 - p)(1 - k)]$$
$$= (1 - u)[1 - 2p(1 - p)k]$$
$$a_{n+1} = u[p^2 + p(1 - p) + \{p(1 - p) + (1 - p)^2\}k'] = u[1 - (1 - p)k']$$

Therefore

$$u_{n+1} = \frac{u[1 - (1 - p)k']}{u[1 - (1 - p)k'] + (1 - u)[1 - 2p(1 - p)k]}$$

$\Delta u = u_{n+1} - u_n$ is positive if

$$u[1 - (1 - p)k'] - u^2[1 - (1 - p)k'] - u(1 - u)[1 - 2p(1 - p)k]$$

is positive, i.e., if

$$1 - (1 - p)k' > 1 - 2p(1 - p)k$$
$$k' < 2pk$$

Thus, as might be expected, which clone is favored depends not only on the selection coefficients but on the frequency of the environments, and the larger the frequency of environment X, the more likely it will pay to canalize for it.

Mendelian Recessive in a Diploid

This is the classical paradigm case. In the Neo-Darwinist formulation, one assumes a fully recessive gene a in frequency u. Then the array of zygotes in generation n is $(1 - u)^2$ AA, $2u(1 - u)$ Aa, and u^2 aa. In generation $n - 1$ this will be changed to $(1 - u)^2$ AA, $2u(1 - u)$ Aa, $u^2(1 - k)$ aa.

In the post-Neo-Darwinist scheme, we have to envisage two environments X and Y in frequencies p and $1 - p$. We can make the same simplifying assumption that both development and selection occur in these environments in proportion to their frequencies. We have to assign selection coefficients to the phenotypes derived from all three genotypes in the different combinations of development and selection. These would be as follows for a case in which the dominant gene produces a fully adaptive development, while the recessive produced canalization for environment X.

	frequency	Developed in X Selected in X	Developed in X Selected in Y	Developed in Y Selected in X	Developed in Y Selected in Y
AA	$(1-u)^2$	1	$1-k$	$1-k$	1
Aa	$2u(1-u)$	1	$1-k$	$1-k$	1
aa	u^2	1	$1-k'$	1	$1-k'$

It is easy to show that the zygotic frequencies in the next generation will be:

AA $(1-u)^2\,[1-2pk(1-p)]$

Aa $2u\,(1-u)\,[1-2pk(1-p)]$

aa $u^2\,[1-k'\,(1-p)]$

Whence

$$u_{n+1}-u_n=\frac{u^2(1-u)\,[(1-k'(1-p)]-[1-2pk(1-p)]}{(1-u)^2[1-2pk(1-p)]+u^2[1-k'(1-p)]}$$

This is positive if k' is less than pk.

Alternatively, one may consider the situation in which *AA* and *Aa* are canalized for environment X, while *aa* produces an adaptive phenotype. The selection coefficients will then be:

	Developed in X Selected in X	Developed in X Selected in Y	Developed in Y Selected in X	Developed in Y Selected in Y
AA and *Aa*	1	$1-k$	1	$1-k$
aa	1	$1-k'$	$1-k'$	1

From this it turns out that $u_{n+1}-u_n$ is positive if $2pk'$ is less than k. Thus, if environment X is the more frequent one (p greater than 0.5), a gene producing canalization (case 1) can make its way against an adaptive gene in the face of a less favorable ratio of selection coefficients than can an adaptive recessive competing with a canalizing dominant.

CONCLUSIONS

It does not seem appropriate to carry the mathematical analysis any further at this time. It has been offered here, not on the grounds that it is immediately applicable or verifiable, any more than is the classical Neo-

Darwinist analysis. What I wished to do was to exhibit a scheme of basic ideas which directs attention towards, rather than away from, the problems which are of most importance for evolutionary theory at the present time. By far the greatest advance in our knowledge of evolution which has occurred in recent years has been the discovery of the enormous range and variety of genetic variation which is present in natural populations. It seems certain that one of the important determinants of this situation is the fact that such populations exist in heterogeneous environments, so that the applied selection criteria are not the same for all individuals. Again, the two major, long-standing problems of evolution are speciation and adaptation. It is generally accepted that a splitting into two taxa involves the operation of two different environments (probably usually allopatric, but possibly in some circumstances sympatric), while an understanding of adaptation demands a theory of how adaptive characters change when the environment changes. In all these three major contexts, the conventional Neo-Darwinist paradigm of the phenotypeless genotype in a uniform environment tends to lead thought away from the challenging questions.

LITERATURE CITED

Cairns Smith, A. G., 1965, "The Origin of Life and the Nature of the Primitive Gene," *J. Theor. Biol.*, 10:53 88.

Haldane, J. B. S., 1924, "The mathematical theory of natural and artificial selection," Part I, *Cambridge Philos. Soc.*, 23:19–41.

-4-

Genetic Control of a Developmental Process

J. M. RENDEL

CSIRO, Division of Animal Genetics
Epping, N.S.W., Australia

Although theorists can base very extensive arguments on assumptions which are few and simple, the nature of the basic assumptions does have a bearing on the direction taken by the argument. Biological theorists commonly make the assumption that both the gene and the phenotype are monotypic units of variation. Fisher (1930), for example, uses human height as an illustration of a phenotype and likens it to fitness. He assumes that maximum fitness, like maximum height, is attainable by directional selection towards an extreme. The gene itself is often represented by some letter, A for example, which mutates to A', the mutation having a more or less direct effect on phenotype, and the steps between the two being safely ignored for the purposes of most arguments.

However, it has been realized for many years that fitness may often, perhaps usually, be at a maximum at some intermediate phenotype about which the population varies; and it is now generally recognized that genes are not all of a kind but can be distinguished by the sort of function they perform in development. Distinctions have been most clearly demonstrated in microbial genetics but must exist elsewhere also. During the past decade I have come to recognize in Drosophila four different kinds of gene which can be distinguished by the sort of function they perform and which co-operate together to carry out a developmental process. It can be shown quite plausibly, I think, that the developmental process is for some purposes the unit in evolution, however much the genes of the four sorts are the ultimate variables; when considering the effects of selection on genes, whether it be natural or artificial selection, it may be of interest to know what sort of gene is concerned. I am going to describe the four types of gene I have come to recognize in Drosophila and some of the evidence for their existence and the way they fit together. The picture is by no means tidy or complete.

The first of the four types of gene which I am going to discuss is the major gene. A major gene is recognized most easily when it mutates to a recessive form. It is a gene which, in addition to having a large effect, has the characteristic of being able to produce a phenotype which is quite precise. This phenotype which is the wild type, being the one commonly found in nature, is the product of a large number of generations of natural selection, and it is not surprising that it is the one out of all the phenotypes associated with the locus which is precisely regulated. By contrast, allelic forms are rare in nature and do not show any particular signs of being adapted to play a specific role in the economy of the organism. The fact that a gene in a diploid has a recessive form implies that the normal wild-type allele is regulated in some way such that one and two doses of it give rise to the same phenotype. Genes whose wild-type allele is dominant though not universal, are common. Omitting lethals and minutes, there are some 365 loci in *Drosophila melanogaster* known to be capable of mutating to a recessive form according to Bridges and Brehme (1944). By comparison, some 80 loci are known only as mutating to dominant alleles; many of these, like Bar and Hairy wing, are not loci in the usual sense, being duplications of one or more bands in the salivary chromosome.

Dominance of wild-type over its recessive allele is an important phenomenon; it is strong evidence of the close control exercised over the normal developmental process, and it is quite different from the dominance of a semi-dominant over its wild-type allele. The former is a dominant in which the phenotype is relatively constant and the heterozygote difficult to tell from the homozygote. The latter is a dominant which is relatively variable, the heterozygote of which is intermediate between the two homozygotes. In the wild-type dominant we are observing a gene which, in the course of evolution, has come to acquire considerable flexibility; this enables it to react to a number of disturbing circumstances in such a way that the phenotype is preserved; one of these disturbing circumstances is the failure of its partner gene. In the mutant dominant, on the other hand, we observe, among other things, what happens when control over the system is broken down. Genes such as Bar and Hairy wing are not closely regulated, particularly in populations into which they have been freshly introduced. No doubt such genes could become regulated in response to appropriate selective breeding, as have certain fancy types, such as crest in poultry. Although in their selected domestic background these behave as wild-type dominants, Fisher (1935) has shown them to be intermediate-type dominants when the mutant gene is transferred to the wild-type background. The presence and absence of horns in ruminants is interesting in connection with the distinction between mutant and wild-type dominance. Taking mammals as a whole, the wild-type condition is hornless; however, the ruminants include a number of families in which the wild-type is horned. In both sheep and cattle the polled condition is dominant to the horned. It has been shown in the New Zealand Romney (Dry, 1955a,b), a breed of sheep which is usually polled, that horns can be produced by two different mutants. One is a good recessive to the polled condition. That is to say, the polled character is the expression of a genotype which still retains a regulation found in most mammals. The other is a good, or reasonably good dominant, showing that the relatively new condition of horn bearing has picked up a large degree of regulation which is still retained by this polled breed of sheep and presumably is present in horned breeds. It is significant that this rare case of a character exhibiting dominance of the wild-type in the mutant form should occur in a character which has a history in which first one and then the other phenotype has been the wild type and so subject to moulding by evolution.

Mather (1949), who makes the distinction between major and poly-genes, has been criticized for doing so on the grounds that the two are not distinct types of entities. In my view he is correct and they are distinct, though I do not believe that they are the only two types which need to be distinguished.

A major gene is a gene usually with large effects, and it is regulated. I suppose it corresponds to Mather's major gene, and that it is a structural gene in the sense used in bacterial genetics.

The developmental process to which I have already referred is related to this major gene. A developmental process is one which is initiated by a particular major gene. Other genes taking part in the process are, in one way or another, hangers-on of the major gene. One of the difficulties to be overcome when studying gene action is that more than one developmental process may contribute to the same phenotype. It is not always easy to dissect out the role played by one major gene and its satellites from the remaining influences. The developmental process with which I shall be concerned most is the one initiated by the scute locus in *Drosophila melanogaster*.

The regulation of the major gene brings us to the genes which do the regulating. I suppose that the major gene is regulated in two ways. One way gives rise to canalization. Genes responsible for canalization see to it that the phenotype does not over-shoot, and they do so by repressing the action of the structural gene when it has done enough. The other kind of regulation is concerned with the extent to which the major gene loci are represented. I can explain best what I mean by citing a familiar example. It is now believed by many geneticists that the sex chromosome in man is never represented in any cell, male or female, by more than one functional chromosome, even when the cell contains three or more X chromosomes. All but one are condensed into the Bar bodies (Lyon, 1966; Greenberg, 1967). Another possible example comes from Callan's (1967) interpretation of the loops of lamp-brush chromosomes as linear replications of the master gene. In the terminology I use, one stage in regulation, (the one dealing with the extent of major gene representation), determines the amount of *make* contributed by the major genes. The second represses this at the appropriate phenotypic level.

I shall deal first with genes which repress the activity of the major gene and reduce it or stop it altogether when the appropriate phenotype is attained. These repressors correspond to the repressors described in microorganisms, whose absence converts an inducible gene into a constitutive one. Many of the structural genes of microorganisms, responsible for the first step in the making of an enzyme, remain inactive until substrate enters the cell. These are inducible genes. The repressor genes which can be shown to be at loci other than the locus of the structural gene can mutate to an inactive form. The effect of having inactive repressor genes is that the structural gene is no longer inhibited when substrate is absent. In the absence of a fully active repressor, the structural gene makes enzyme all the time regardless of the presence or absence of substrate and is called constitutive. It can be shown by making heterozygous diploids of various kinds that a normal structural gene is repressed by a normal repressor gene. An appropriate mutation of either repressor or structural gene can destroy repression and result in a structural gene which continues to pour out its product regardless, the structural gene being insensitive to repressor substance or the repressor gene making an incompetent supressor substance or none at all.

I have assumed that the major gene of higher organisms is an inducible structural gene. It is repressed by a repressor and induced, one must suppose, by some endogenous substance; possibly a substance present in the original egg cytoplasm or nucleoplasm, or possibly by a substance which is made by another gene during the course of development. What the relationship of the repressor is to the inducer I have no idea; the repressor substance of the scute gene appears to be made by two or more genes not at the structural gene locus in response to end product of the scute developmental process as a whole. The evidence that there is repression and that it is of the major gene by a substance made by another genotype in response to end product, comes in higher organisms from experiments on selection, and must be looked at next.

A number of selection experiments have shown that when selection is imposed upon a population homozygous for a recessive gene, the recessive phenotype can be made to approach wild type; but once it reaches wild-type, the phenotype does not increase further. In the population as a whole, change can be shown to be going on insofar as a smaller and smaller fraction of the population has a mutant phenotype, and those individuals that do, have phenotypes which are closer and closer to wild-type; but though a larger and larger proportion come to have wild-type phenotypes, none exceeds wild type. Figure 1 is a series of histograms showing different stages in a selection experiment. Class four is the wild-type phenotype and is, as a matter of fact, the number of scutellar bristles in *Drosophila melanogaster;* however, this illustration is typical of what happens when one selects any recessive towards wild-type. It is a common observation and was discussed in the early days of genetics under the heading of expressivity and penetrance, the well-known correlation between these two showing that the fewer the abnormal individuals the less will their abnormalities be. There is ample evidence of this sort showing a ceiling to phenotype; a ceiling is evidence suggesting that phenotype is controlled—that once it reaches wild type, it is not allowed to over-shoot.

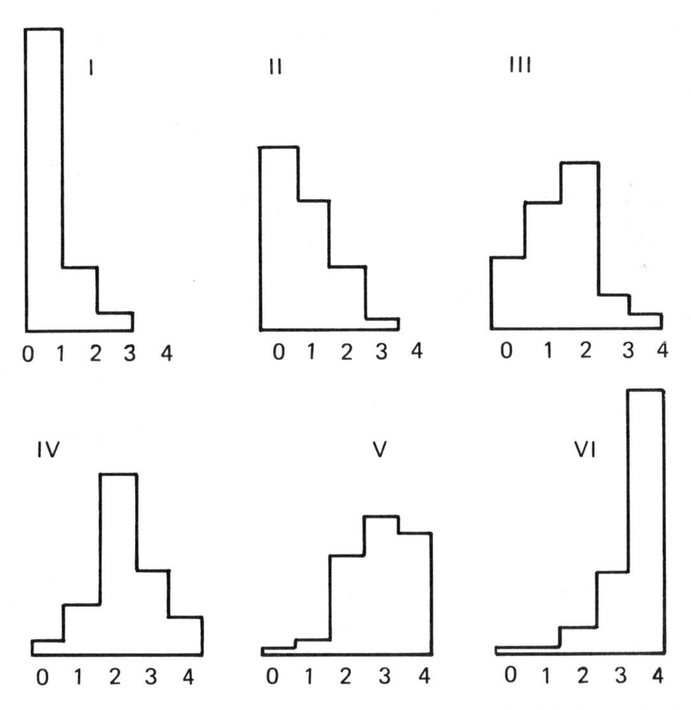

Figure 1. I–VI are size stages in selecting for increased scutellar bristle number in a scute stock. The histograms show the frequency of this with 0 to 4 bristles.

We have next to show that control of the phenotype is exercised by repression of the major gene. One argument is that control is stronger in the presence of a strong gene than a weak one, and the first step in this direction is to show how strength of control can be measured. Measurement depends on the fact that you can over-shoot the wild-type phenotype if you persevere; therefore, control over the phenotype is a limited control. This has been shown for a number of characters, e.g., scute in Drosophila (Rendel, 1959), shaven in Drosophila (Schultz, 1935), Tabby in mice (Dun and Fraser, 1958), and sterile base in wheat. In all these, the wild-type phenotype can be transcended by selection, which brings the expression of the recessive mutant genotype towards normal. This can be illustrated by looking at some more scute stocks. The phenotype we are concerned with is that of the number of bristles on the scutellum. Some flies are shown in Figure 2. The figure shows flies which are wild-type at the major gene locus in the second row and flies which carry sc^1 in the first row. Typical

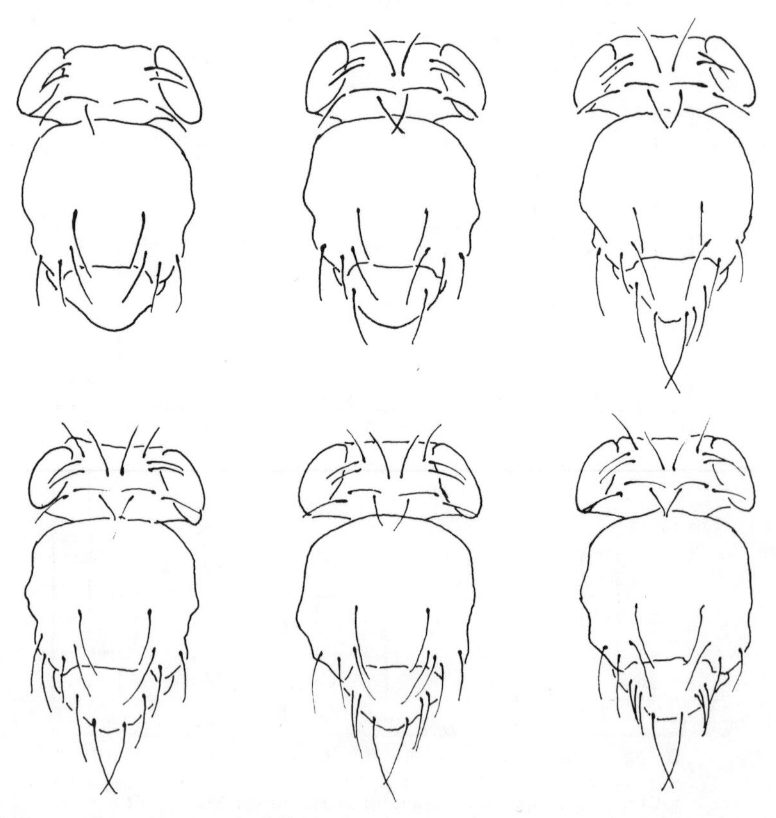

Figure 2. Top row: *sc sc* flies; bottom row: + + flies. Selected for high bristle number on the right.

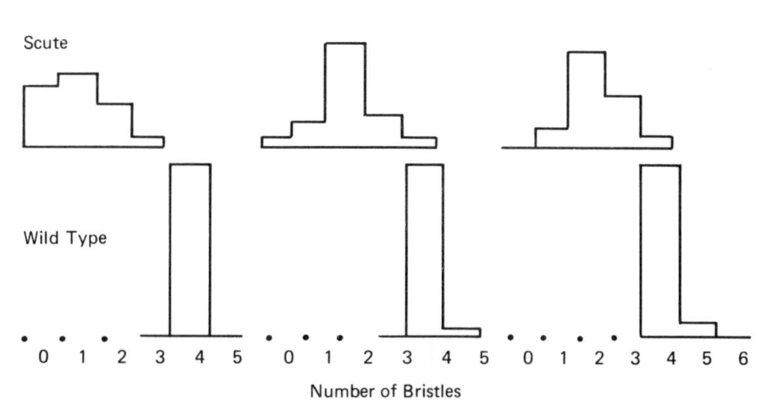

Figure 3. Distribution of bristle numbers at three stages of selection for bristles in a population segregating for scute.

flies are to the left; results of selection to the right. Figure 3 shows what happens when one selects for high bristle number in scute flies in a population segregating for wild type and scute. Three stages in selection are shown, and it can be seen that selection in scute flies is eventually effective on their wild-type sibs. Selection is carried out on the scute flies, because it is thought that this will pick out genotypes which increase bristle number by reinforcing the deficiencies of the mutant gene. Throughout the thesis I am trying to develop here, I am referring to genes which belong to a single developmental process. Genes have been selected for their effect on the phenotype in the presence of the mutant gene in order to include, as far as possible, only genes of one developmental process and not all and every gene which increases the phenotype. The effect of selection in scute on wild-type sibs is the same as the effect on scute flies if both are measured in probits. Another way of expressing the effect of selection is to plot increase in mean phenotype in actual bristles against progress measured in probits. This is shown in Figure 4. It is a better method of illustrating the plateau or platform at four bristles.

Having shown that the ceiling at four bristles is not a ceiling so much as a platform which can eventually be crossed, the next step is to measure the width of this platform. If this platform is to be interpreted as due to repression of the major gene, its extent or width will be important, for when all the major gene that can be repressed has been repressed, there will be nothing left with which to control the phenotype, and we may expect to find the width of the platform proportional to the strength of the major gene. It has been found convenient to measure the width of the platform from the fraction of a population which lies outside it: in the case of scute, by counting the fraction of flies having more or fewer than four scutellar bristles. The proportion of the population lying above three and below five

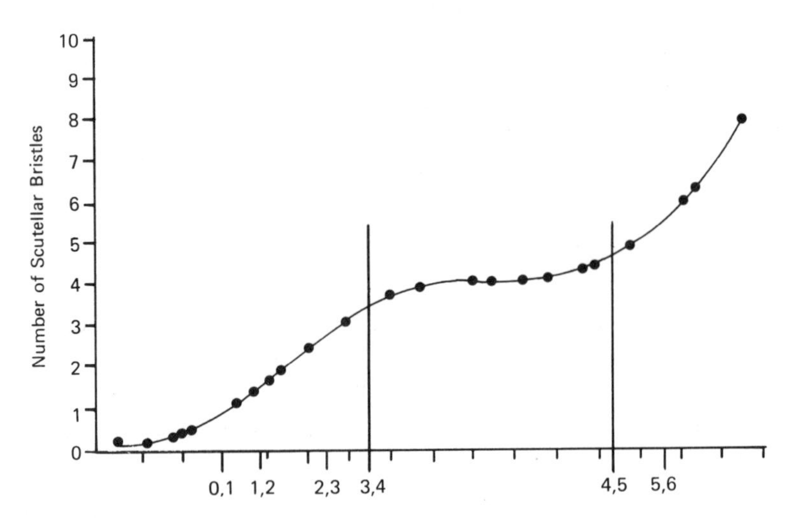

Figure 4. Probit scale of scutellar *make* values showing bristle thresholds (1 div = 1 standard deviation of *make*).

is expressed in probits to offset any asymmetry which would be introduced if one had a relatively large number of flies with three or five bristles. The number of standard deviations spanned by the 4-class in this case is taken as a measure of the strength of repression.

With this measure of the strength of repression, we can test the difference in degree of control over the scute phenotype in homozygous scute and homozygous wild type. In homozygous scute regulation is by suppression of the scute gene; in homozygous wild type it is by suppression of the more powerful wild-type gene. In wild type, the 4-class spans some 5.5σ to 6.0σ; however, in scute homozygotes which have been selected up until a fraction have 5 bristles on the scutellum the width of the 4-class is only 3.5σ. The fact that the width of the 4-class is less in scute flies is compatible with the view that control is by repression of the gene at the major gene locus, and suggestive of it but does not prove it. According to this hypothesis, the wild-type allele has a range of potential but repressible activity equivalent in phenotype to 5.5σ. Whereas in scute this range is only 3.5σ, in both genotypes there is some activity of the major gene which is not repressible, and both major genes are supported by minor gene activity.

Figure 5 is a diagram which represents a scute gene and a wild-type gene acting in a high and a low background of minor genes. The vertical scale is in standard deviations of bristle number. The blocks represent bits of the gene which are repressible. The unfilled blocks have been repressed; the filled ones have not. The unfilled ones can be called on in need, and should the minor gene activity fall, bringing them below the 4/3 line, they would be called on. The straight lines represent an amount of phenotypic effect

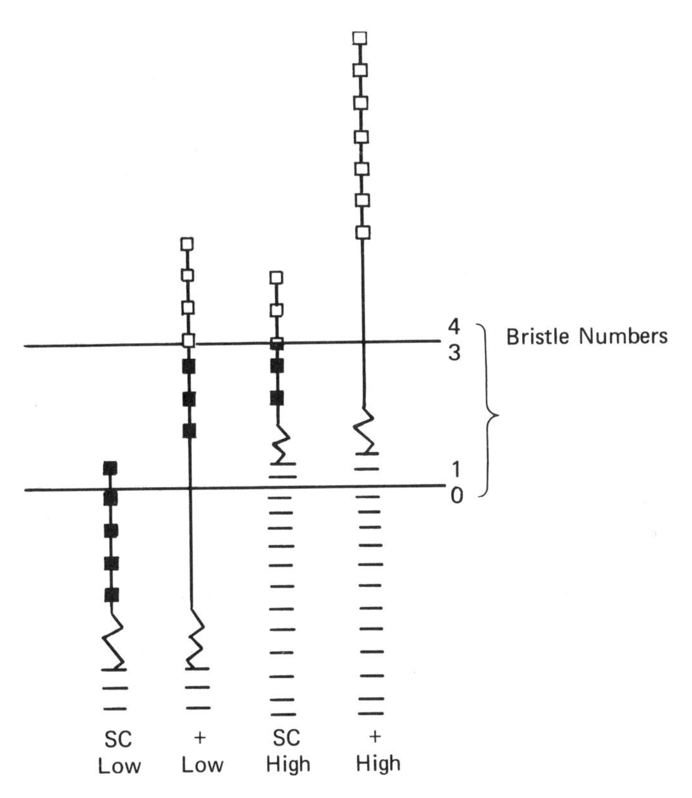

Figure 5. Relative strength of repressible and nonrepressible gene potential in *sc* and +.

that cannot be repressed and is irrevocably added by the gene. In the high line, the *sc* gene is shown with two filled and two unfilled blocks straddling the 3/4 bristle boundary. This means in practice that in this line the majority of scute flies has four bristles. About four per cent have five or more and about four per cent have three or less, the top unfilled block being 1.75σ above the 3/4 cut-off and the bottom filled one being 1.75s below it. The fraction of a population lying further than 1.75σ from the mean is about four per cent. The performance of scute in the high line tells us that the activity of the *sc* gene can be repressed to the extent of 3.5σ. The scute gene in the low line never reaches a phenotype at which 4 is repressed. All 3.5σ, represented by filled blocks, and an unknown further amount represented as a zigzag is not repressed; below this again is shown the activity of minor genes which is also of an unknown extent. The difference between *sc* in the high line and *sc* in the low line is what has been added by minor genes. This addition is therefore measurable. The + flies in the low line have the same complement of minor genes as the *sc* ones, and the difference between them and the *sc* flies is due to the major gene. Part of this difference is due to extra repressible activity; there is in addition an amount equivalent to 5σ which is not repressible. In the high line, this unrepressible bit takes the mean well above four bristles, all the repressible fraction being repressed all the time. So in *sc* flies, there is an unknown quantity of non-repressible gene activity represented as a zig-zag, in + there is 5σ more than this. Supposing that the nonrepressible is proportional to the repressible, which might be the case if the *sc* gene made a less active enzyme than the + gene. We could estimate the unrepressible activity since in + it would be $\frac{5.5}{3.5}$ what it is in *sc*. In *sc* it is u, while in + it is $u + 5\sigma$. Since $u = 8.75$, the total activity of a + gene becomes 12.75 unrepressible and 5.5 repressible, and of a scute gene, 8.75 and 3.5. The suggestion that control is through repression of the major gene comes also from a similar line of argument leading out of a demonstration that the control of the phenotype is by a genotype at some locus other than the major gene locus. If we follow the model provided by microorganisms and suppose control is exercised by repression of the major gene, then it should be possible to select for genes which will repress a major gene locus not already controlled at a precise phenotype, and to produce out of an uncontrolled phenotype with no dominance one which is controlled and showing dominance of the sort associated with the wild type (Rendel *et al.* 1960, Rendel *et al.* 1966). As an uncontrolled major gene, one can use any recessive allele of a major gene, and in the case of the scute locus a population homozygous for the mutant *sc'* was chosen and selected to have precisely two bristles. Males were selected with one or two bristles as was needed to keep the mean of the population at two; females all had exactly two, and both males and females were selected from a sibship in which there was minimal variation about two. Figure 6 shows the progress of selection. The 2-class increased in width from about 1.4σ to 3.5σ in two steps during 50 generations of selection, each step occupied 15 generations or so, and the succeeding 180 generations of selection have added a further 1σ so that now the 2-class in females occupies about 4.5σ. In other words, in this population over 97 per cent of females have two scutellar bristles. The strength of canalization in males is rather less.

It has been shown that the gene or genes responsible for this constancy of phenotype reside mostly on the third, partly on the first, and perhaps a little on the second chromosome. It is interesting to notice that the gene at the scute locus in this stock has gained in power. The first stage in selecting for its repression at a phenotype of two bristles led to a width of control of 3.5σ; this is exactly the same as the width of control of the original scute allele at four bristles, and reinforces the suggestion that control of the

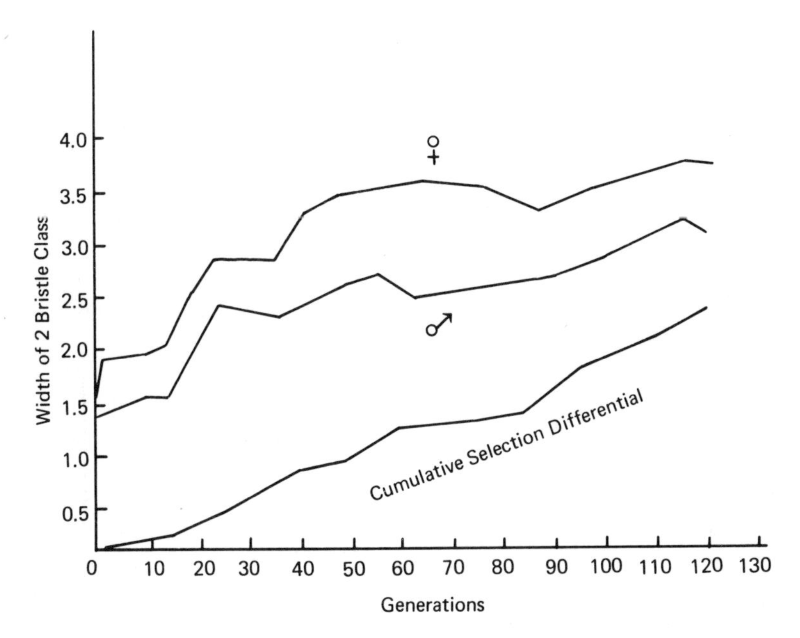

Figure 6. Response to selection for constancy at two bristles in ♂♂ and ♀♀.

phenotype is proportional to the strength of the major gene and brought about by its repression. In later generations, selection for the canlization at two bristles has increased the width of the 2-class still further, and we should expect that at the same time the gene at the scute locus has become more powerful. That it has done so is demonstrated by backcrossing the original sc^1 allele marked by white into the low variance stock marked by blood. The counts are shown in Table I. There are two things of interest

TABLE I

Back Cross	*sc w* ♂				*sc w^bl* ♂				Difference in σ
	0	1	2	3	0	1	2	3	
1	21	25	13	-	5	9	15	1	0.82
2	39	59	56	1	14	47	112	2	0.97
3	26	73	120	1	8	37	201	4	0.93
4	4	27	70	-	2	14	91	3	0.67
	sc w^bl/sc w^bl ♀				*sc w^bl/sc w* ♀				Difference
	0	1	2	3	0	1	2	3	
1	-	2	38	2	-	3	37	1	0.27
2	3	5	133	9	2	15	134	5	0.35
3	-	8	182	11	1	9	217	3	0.33
4	-	1	76	1	-	1	102	1	0.01

in this count: first, the difference between the white and blood males, and second, the absence of any significant difference on the phenotypic scale in the females. However, on the probit scale measuring from the 2/3 and the 3/4 cut-off and taking a mean of the differences, it can be shown that the heterozygous females differ from the homozygous ones by about ¼ to ⅓ the difference between males. Thus we have dominance, but it is achieved in the face of a detectable difference between the two classes of female on the one hand, and between males and females on the other. In this respect it parallels the dominance of + over *sc.* The fact that canalization was achieved by some 50 generations of selection in two phases of 15 generations each, suggests it is due to several genes: the fact that the canalizing effect can be located to two possibly three chromosomes out of four shows it is due to at least two genes. Our finding that the width of the 2-class in early generations of the low variance stock is the same as the width of the 4-class in high selection scute stocks supports the ideas that suppression of the major gene is responsible. The fact that this width increases in later generations and the increase is accompanied by a greater power of action of sc^1 strengthens the idea still further. If canalization is by repression of the major gene, presumably the genes responsible for repression made a repressor and do so when activated by an inducer related to the final phenotype. We may refer to this for simplicity as end-product. It is to be noted that the nature of the major gene does not determine the phenotype at which repression takes place. In ordinary stocks, both the scute gene and the + gene are repressed at four scutellar bristles. But selection in the low variance stock has introduced additional suppression of the *sc* gene at a phenotype of two bristles; and when scute in low variance back mutates to wild-type, which it does with great frequency, the + phenotype is seen to be still closely regulated at four bristles. We may infer from this that the repressor genes are activated by the product of the developmental process as a whole, not simply by the product of the major gene. It is difficult to imagine that repressor genes would respond to more than one inducer. It follows that those bits of the developmental process which are not due to the activity of the major gene either give rise to the same end product by a different route or act through the major gene, increasing its output. No doubt both can occur. This brings us to a third class of gene, the minor gene.

The minor genes are those genes with small effects which are accumulated as soon as one starts selecting to change a phenotype. Of course, if the major gene mutates to a different allele, this change also will be subject to selection. But in general, what is typical of directional selection is the accumulation of many minor effects brought about by a number of minor genes with small effect. Minor genes are best demonstrated by selection. As one might suppose, they are not repressed, or if they are, it is not with respect to the phenotype for which they are the minor genes. The best evidence that minor genes are not regulated can be explained with reference to Figure 5. When, as in a high line, all regulation of the major gene is used up, there can be no adjustment for fluctuations in the environment and for the effects of segregation of minor genes, and it is at just this point when expression of the character is abnormally high that one finds variation about the mean at its maximum. Furthermore differences between selection lines due to minor genes increase to a maximum when the bristle number in the lines is due to increased addition of extra major genes. If minor genes were repressed in response to over-production, one would expect their effects to be minimal and their variation minimal at higher levels of expression. I should point out here that selection which is of a type to encourage extra bristles by some route other than the scute developmental process will complicate the picture. Scutellar bristle number in wild-type can be increased in a number of different ways which are not necessarily concerned with the scute developmental process. Fraser and Green (1964) have shown, for example, that selection for extra scutellars in wild type is accompanied by some increase in bristle number in scute, but not as much as one would expect from the increase in wild type. Rendel, Sheldon, and Finlay (1965) confirm this. Further, if several lines are taken, the relationship between the change in scute and in wild-type flies is different in different lines. In contrast, when selecting in the presence of scute, both Dr. Sheldon and I have found that modifiers of *sc* act equally on wild-type.

One important question about minor genes is whether they modify the action of the major gene or supplement it. If a deletion, for example, could be modified, it would suggest that the minor genes act on their own to supplement the major gene. If, on the other hand, the minor genes were inactive in the presence of a deletion of the major gene, it would suggest they act by multiplying what the major gene does. A *y ac sc pn* chromosome was made by Muller in such a way that it was expected to contain a deficiency for *sc*. It certainly has very low expression; in unselected stocks no scutellar bristles are found. However, it is modifiable by the background genotype of a scute up-selection line. In this background, flies carrying the deletion in homozygous condition do occasionally produce scute bristles. Even if this chromosome does not carry a deficiency, the experiment suggests that the minor genes of the scute high line act as supplements and not multipliers, for they add about the same to the deficiency as they do to the scute′ allele, and certainly no less, which they would be expected to do if they were modifiers of the action of the scute gene. Hubby and Lewontin (1966) have shown by gel electrophoresis that enzymes in Drosophila are distributed in a way which at first sight seems to parallel the action of a major gene and its satellite minor genes. The esterases they looked at, for example, are made mostly by one locus which is known to mutate to at least five forms. Its esterase appears as a major band, or in heterozygous sets of major bands. In addition, there are nine other sources of esterase. These appear as very faint bands, often only visible in bulked squashes and appear to be made by independent loci, though I think the number of loci is not absolutely proved as yet, nor is it absolutely proved that the esterases are at all interchangeable in metabolism although they do all split the same substrate *in vitro*. If it should turn out that Hubby and Lewontin are dealing with a major gene and its satellite minor genes, they will have proved that the minor genes make some product with effectively the same function as the product of the major gene, but less of it and in a less effective form.

If the enzymes showing as minor bands supplement the effect of the enzyme showing as the major band, selection for increased esterase activity at the time of assay chosen by Hubby and Lewontin might be effective in a number of ways. A number of different esterases will be manufactured during development with different specificities and different times of appearance. Insofar as they have any action on substrate for which they are not specific, selection may well move their time of appearance somewhat, and also their specificity, and so make them a little more effective as modifiers. At the same time, if they have a major role to play in another developmental process, any regulation to which they are subject may be expected to take place with respect to this process which will almost certainly take place at another time and in another context.

To conclude, minor genes have small effects; they show no marked dominance and are not regulated, at least they are not regulated with respect to the developmental process in which they act as minor genes. They act to supplement a major gene, bringing it to within easy striking distance of its target.

The fourth class of gene is responsible for the extent to which the major gene takes part in the reaction. By this I do not mean the degree to which it is regulated; I mean the amount of it which is present and the amount which *can* be regulated. I refer to the amount of *make* it contributes to the developmental process. There are a number of ways in which one might imagine this happening. Callan has suggested that a loop in a lampbrush chromosome is composed of a number of slave genes made by the master gene in its own image. If this were so, the number of replicates would fix the extent to which the major gene was represented and was capable of taking part in a reaction. Lyon has suggested that the whole of one X in a mammal is out of action from a certain stage of development on. In this case, whereas in early stages of development each gene is represented twice, in later stages its representation is reduced to once in each cell. It is possible that genes at all loci can be cut out in this way, though this seems very improbable in Drosophila from evidence of eye color heterozygotes. I have suggested elsewhere that the magnitude of the role played by the major gene may have its upper limit set by the number of ribosomes it can collar for its own use: presumably major genes would be in competition with minor genes and with each other for ribosomes. The point is that some mechanism fixes the extent to which the major gene enters into the developmental process. The major gene is adjusted to the number and strength of the minor genes in such a way that the mean value of the character is fixed at an appropriate level in unselected normal stocks. The evidence for adjustment of representation of the major gene to match the minor genes comes partly from gene dosage experiments and partly from a measurement of the extent to which the scute character is regulatable in different selection lines (Rendel *et al.* 1965) If regulation acts through the major gene, the more the major gene is represented the more extensive one would expect regulation to be because there is more major gene to play with. Consider first the homozygous wild-type females and hemizygous wild-type male. The amount of major gene activity in males and females is about the same. This is in spite of the females carrying two genes to the males' one. Furthermore, the degree of canalization is the same in both sexes except that in males the width of the 4-class is usually half a standard deviation greater than in females. This suggests that in males and females the amount of suppressible gene action is about the same in the two sexes, perhaps a little more in males. In males it appears that the same total activity may be made up of a higher proportion of suppressible gene activity. If one now adds an extra wild locus to each so that males carry two scute loci and females three, very little is added to gene activity in a high line but some considerable amount in a low line.

There appears to be some irreducible addition which is about 1 to 2σ. But the addition is made at the expense of repressible gene activity. At the same time that the total gene activity is increased, the fraction of it which is repressible is reduced. This addition of 1 to 2σ to total activity represents an addition of about twice this amount to irrepressible gene activity and a subtraction of 1 to 2σ from repressible gene activity. It seems as though the major gene locus were competing for representation with minor gene, and that with the addition of extra major genes, the non-repressible additions compete with the repressible fraction. The behavior of the major gene in this respect is independent of phenotypic expression as can be seen by the exact parallel behavior of the loci carrying the scute gene. At a lower phenotype they follow the same course as wild-type. The scute gene, whether represented as + or sc^1, has to compete with other genes for opportunity to act, its success being proportional to the number of loci competing, not the activity of the gene at the locus (Figure 7). The outcome in phenotypic terms is, of course, affected by the effectiveness of the genotype which becomes established. If the extent of representation of the major gene is determined by something in the nature of an elimination of one gene, partial or complete as in the Lyon hypothesis or as a duplication of the gene as in Callan's explanation of lampbrush chromosomes, it is possible that the outcome will depend on other genes and selection may be able to influence it. Two experiments may bear upon this point. In one, parents were selected from cultures in which *Hw* + females were as far as possible the same in expression as *HwHw* ♀♀. Selection took place in two lines. In one line, selection from the cultures in which heterozygotes most nearly resembled homozygotes was at random; in the other it was aimed to keep the mean score constant. In practice, selection in the first line was for *Hw*+ females which had a relatively high expression, whereas in the latter, selection was for females which had a relatively low expression. The results are shown in Figures 8 and 9 (the figures are taken from unpublished experiments to be submitted as part of a thesis for a Ph.D. at Sydney University by Ohh Bong Kug). In both lines, the heterozygote has come to approach the homozygote. In the line in which the mean was not

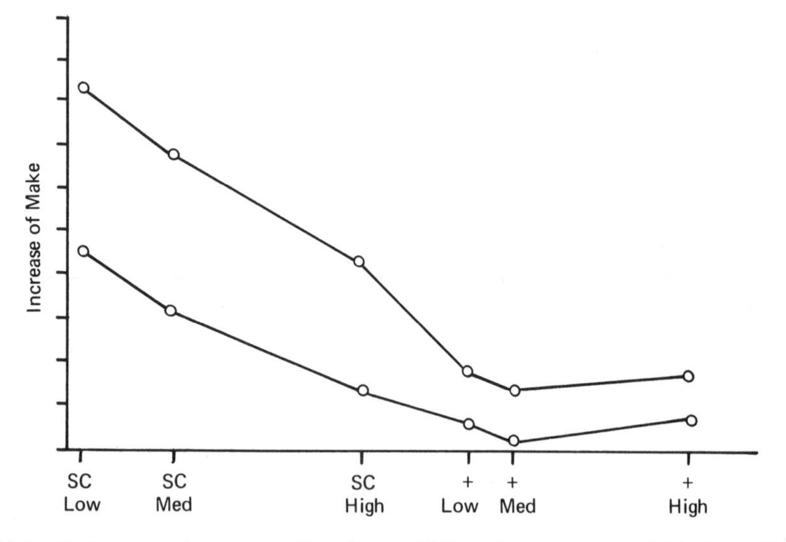

Figure 7. Increase of *make* resulting from addition of a *sc* or + to the last genotype. Upper curve: + gene added on Y chromosome. Lower curve: *sc* gene added on Y chromsome. Ordinate shows increase in *make*, abscissa shows relative levels of *make* of different host genotypes.

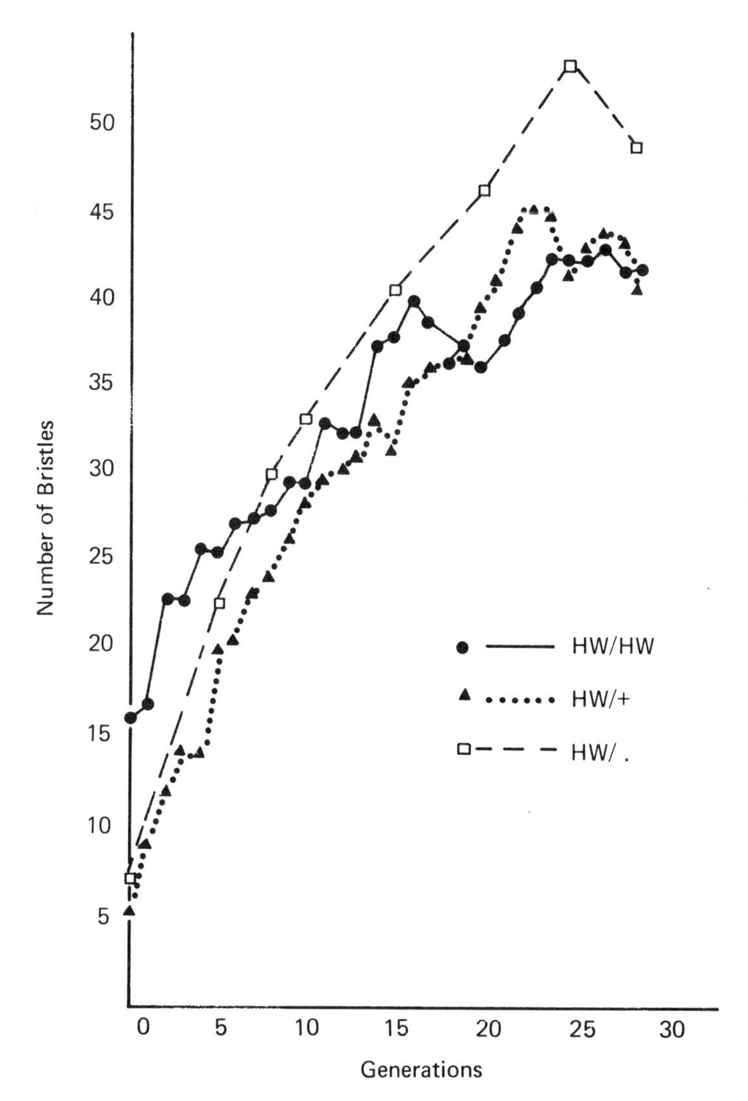

Figure 8. Selecting for equality of expression of *HwHw*, and *Hw+*, selecting for *Hw+* near to or equal to *HwHw*.

kept down, bristle number has shot up. Strangely enough, males have outstripped both classes of females. In the line in which mean bristle number was kept down, the mean of both classes of female score the same as the original heterozygote score and variance about this score has been halved, most of the reduction being due to an increase in the width of the mean class which has doubled while the width of other classes has stayed the

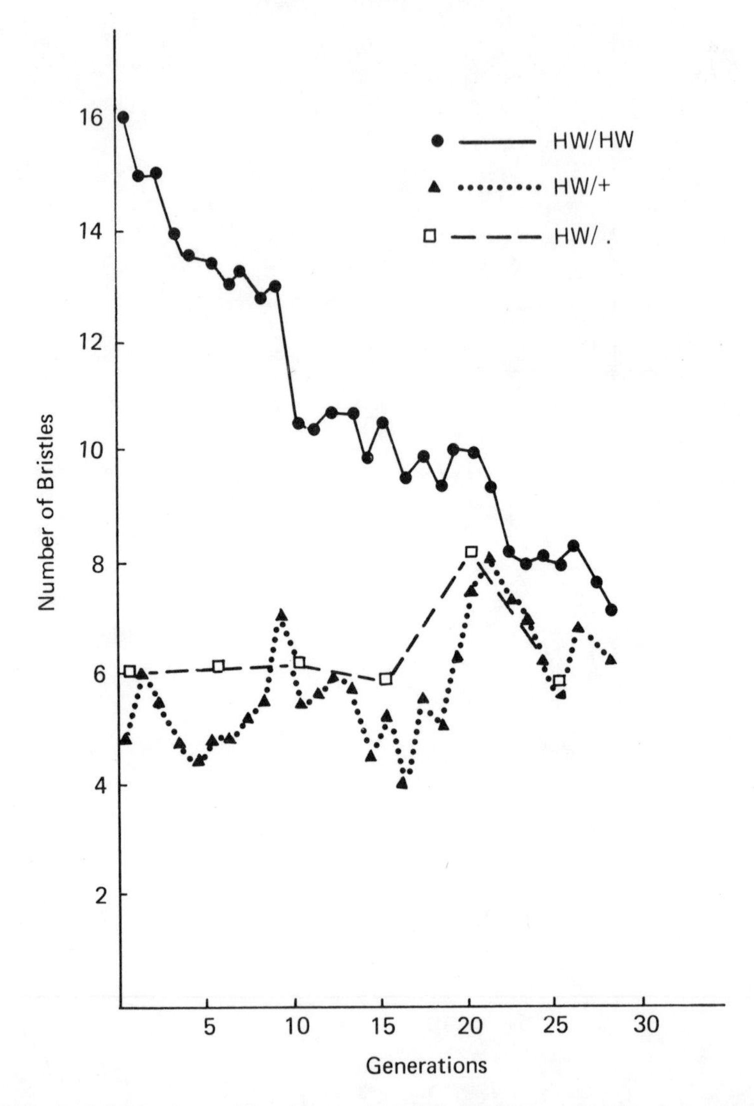

Figure 9. Selection for equality of expression of *HwHw* or *Hw+*, keeping the mean at 6.

same. In the second experiment, abdominal bristle number on the last segment in sc^1 flies was the character selected. In males this averages about 5.5; in females 9. Selection was for males and females with 7. In the fortieth generation the mean of males was 6.1 and of females 7.2, so there has been a considerable approximation. There has been no obvious increase in the width of any class measured in probits. It is not easy to interpret these experiments as they stand. The second experiment is a direct demonstration that a genotype can be selected which gives males and females the same expression. If we are dealing with the extent to which the major gene is represented, the *Hw* experiment suggests that selection can favor the presence of one allele over the other. Whatever the explanation, equality of expression has been achieved without canalization. This is to be contrasted with selection for canalization of scutellar bristle number at two which has been achieved withoug equalizing the two sexes. There are thus two independent phenomena observed: one controls expression at a given level, the other equates *make* in homo- and heterozygotes, in males and females. I suppose there is a genotype which brings this about because it has been achieved by selection.

In summary, a typical developmental process is initiated by the induction of a major gene. This gene, either by replicating itself or competing with minor genes for ribosomes or by some other means under genetic control, establishes itself in competition with a number of minor genes which bring the whole reaction to within easy reach of the preferred phenotype. Part of the *make* attributable to the major gene is repressible and is repressed by a genotype which is itself induced when the right phenotype is attained. The minor genes are not repressed, and so the major gene is the sole source of regulation and is repressed by a repressor genotype which is itself induced by the end product of both major and minor genes.

This organization is reflected in the phenotype whose variation, when plotted against *make,* follows a sigmoid curve and exhibits interactions, both dominant and epistatic, which one expects from such a system.

LITERATURE CITED

Bridges Calvin B., and Katherine S. Brehme, 1944, "The Mutants of *Drosophila melanogaster*," *Carnegie Inst. Wash. Pub.*, 552.

Callan, H. G., 1967, "The Organization of Genetic Units in Chromosomes," *J. Cell Sci.* 2:1–7.

Dry, F. W., 1955a, The Dominant N Gene in New Zealand Romney Sheep. *Australian J. Agr. Res.*, 6:725–769.

———, 1955b, "The Recessive N Gene in New Zealand Romney Sheep," *Australian J. Agri. Res.*, 6:833–862.

Dun, R. B., and A. S. Fraser, 1958. "Selection for an Invariant Character, Vibrissae Number, in the House Mouse," *Nature* 181:1018–1019.

Fisher, R. A., 1930, *The Genetical Theory of Natural Selection*, Clarendon Press, Oxford.

———, 1935, "Dominance in Poultry," *PhilTrans Roy. Soc.*, B 225:197–226.

Fraser, Alex, 1963, "Variation of Scutellar Bristles in Drosophila. I. Genetic Linkage." *Genetics*, 48:497–514.

Fraser, Alex, and M. M. Green, 1964, "Variation of Scutellar Bristles in Drosophila. III. Sex Dimorphism," *Genetics* 50:351–362.

Gruneberg, H. G., 1967, "Sex linked Genes in Man and the Lyon Hypothesis," *Ann. Hum. Genet.*, London, 30:239–257.

Hubby, J. L., and R. C. Lewontin, 1966, "A Molecular Approach to the Study of Genic Heterozygosity in Natural Populations. I. The Number of Alleles at Different Loci in *Drosophila pseudoobscura,*" *Genetics*, 54:577–594.

Lyon, Mary F., 1966, "Chromosome Inactivation in Mammals," *Adv. Teratol.* 1:25–48.

Mather, K., 1949, *Biometrical Genetics—The Study of Continuous Variation*, Dover Publications Inc., New York.

Rendel, J. M., 1959, "Canalization of the Scute Phenotype of Drosophila," *Evolution*, 13: 425–439.

Rendel, J. M., and B. L. Sheldon, 1960, "Selection for Canalization of the Scute Phenotype in *Drosophila melanogaster,*" *Australian J. Biolog. Sci.*, 13:36–47.

Rendel, J. M., B. L. Sheldon, and D. E. Finlay, 1965, "Canalization of Development of Scutellar Bristles in Drosophila by Control of the Scute Locus," *Genetics*, 52:1137: 1151.

______, 1966, "Selection for Canalization of the Scute Phenotype. II," *Amer. Natur.*, 100:13-31.

Schultz, Jack, 1935, "Aspects of the Relation Between Genes and Development in Drosophila," *Amer. Natur.*, 69:30–54.

-5-

Evolutionary Consequences of Flexibility

RICHARD LEVINS
University of Chicago
Chicago, Illinois

In these remarks I will make two assertions, two distinctions, introduce two models, and demonstrate six consequences. *Assertion I* (the optimization principle): Populations or species will, in general, differ in the same direction as their optima differ where the optimum is defined as the maximum value of the appropriate fitness function which satisfies certain constraints.

Optimization methods will be used by way of fitness sets which will be introduced here for two environments but are readily extended to any number. *Model I*: Each phenotype available to a population can be represented by a point on a graph whose axes are fitnesses in two alternative environments. The set of all such points constitutes a fitness set. In the absence of any special interaction between unlike types, the fitness point of a mixed population will lie on the straight line joining the fitness points of the components. *Model II*: If instead of using phenotypes we use genotypes, the two alleles at a single locus give rise to three fitness points, while all possible populations lie on a curve joining the homozygote fitness points and bending toward the heterozygote. Most of our results apply to both models although they will be shown only for one.

Distinction I: We distinguish between fitness sets which are convex and concave along their upper right-hand boundary. If the fitness set is convex, each mixed population of Model I is inferior in both environments to some monomorphic population, so that all optima are monomorphic. This, of course, is not true for concave fitness sets. It has been argued elsewhere and is here offered as *Assertion* II: that if two environments differ little compared to the tolerance of the individual, the fitness set is convex, while if they are very different, the fitness set is at least partly concave. We now indicate that phenotypic flexibility increases convexity. In Figure 1 we show

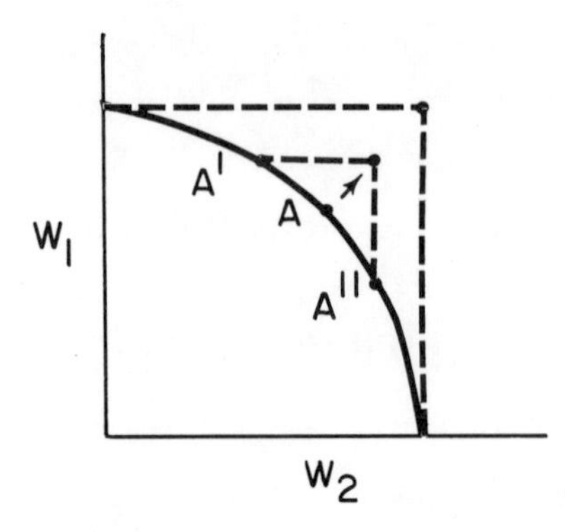

Figure 1. Flexibility increases the convexity of the set. Point A becomes A' for environment 1 and A'' for environment 2.

an arbitrary fitness set. Suppose that the genotype giving rise to point A is flexible. It responds to some environmental signal in such a way that in environment I the phenotype is displaced to A' and in environment II to A''. The net effect is to displace A up and to the right. The greater the flexibility, the greater this effect, and in the limit, the fitness set would be the rectangle shown by the dotted line. Therefore, for purposes of further discussion, increasing flexibility will be identified with increasing convexity of the fitness set.

The optimum population is the one which maximizes some adaptive function $A(W_1, W_2)$, an increasing function of the fitnesses W_1 and W_2 in the two environments. We now make *Distinction II*, between coarse-grained and fine-grained environments. In a fine-grained environment, the individual passes through or utilizes many small units of environment. Thus, the environment affects it as an average, and the appropriate function to maximize is

$$A(W_1, W_2) = pW_1 + (1 - p)W_2 \tag{1}$$

where p is the frequency of environment I. A coarse-grained environment faces the individual as alternatives. The whole life is spent in a single patch, or the environmental fluctuations last a whole generation. Then the function to maximize is

$$A(W_1, W_2) = W_1^p W_2^{1-p} \text{ or } p \log W_1 + (1 - p) \log W_2. \tag{2}$$

As the environment becomes increasingly coarse-grained, the adaptive function bends from the straight lines of equation (1) to the hyperbola-looking curves of equation (2). In either case, the optimum is the point on the fitness set tangent to the adaptive function $A(W_1, W_2) = K$ for the largest K.

We can now apply these techniques to derive the following:

1. On a convex fitness set, as p changes, the slope of the adaptive function changes and the point of tangency moves to a point on the fitness set with the same slope. The more convex the fitness set, the greater the change of slope per unit arc. Therefore, a smaller change in the optimum population occurs with a given change of environment. Hence, the more flexible the individual, the smaller will be the temporal fluctuations. In Model II, the more convex the fitness set, the more stable will be the polymorphism.

2. The same argument applies to a spatial distribution: the smaller the flexibility of the individual the steeper will be the cline in response to an environmental gradient. In the gene frequency terms of Model II, flexible species will have widespread polymorphisms with little change in frequency while inflexible species will have local polymorphisms varying in frequency and replaced from region to region.

3. In Model I, the fitness set is vertical and horizontal at the ends. But in Model II this is not true. Here the slope begins at some initial value m_1 and ends at m_2.

Only when both environments are sufficiently abundant for the slope of the adaptive function to fall between m_1 and m_2 will there be polymorphism. Otherwise, the homozygote will be fixed and the species will be specialized. The interval m_1, m_2 is narrow for a flat convex fitness set and increases with convexity. Therefore, the more flexible the species, the more likely it is to take evolutionary notice of even rare environments while inflexible species will tend to be more specialized, to have narrow niches.

4. Fine-grained selection on a concave fitness set results in a monomorphic specialized population. As seen from Figure 2, although one specialized monomorphic alternative is usually better than the other, both may be better than the intermediate phenotypes. Therefore, either specialization is stable. This may lead to (a) discrete races, (b) a pattern deter-

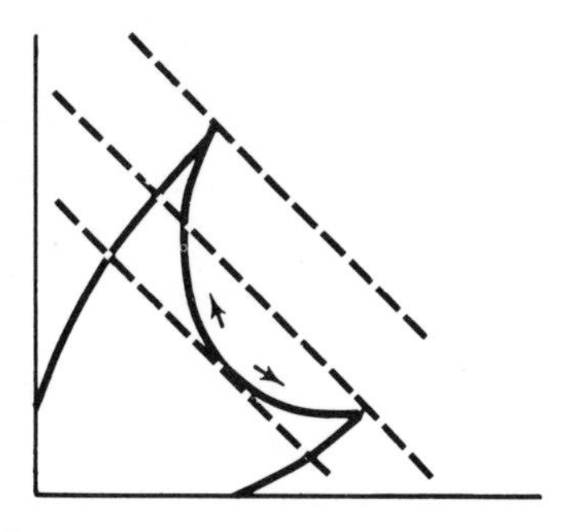

Figure 2. On a concave fitness set with fine-grained selection, there may be two optima. Phenotype will evolve toward one or the other depending on the starting position.

mined by past history (which specialization was adopted in the past), and (c) a barrier to spread. When different environments are still similar enough to give a convex fitness set, a population can adapt in a continuous fashion to changing proportions of them. But an environmental difference great enough to give a concave fitness set will be a barrier. Hence, a species can evolve to meet environmental differences that are smaller than the tolerance range of the individual. Flexible species can, therefore, be more widespread and survive longer. The same argument applies to coarse-grained selection if the environmental difference is great enough. Otherwise polymorphism will result.

5. As environments become less similar (or equivalently, when flexibility decreases), an originally convex fitness set becomes concave in a small part of the boundary which then increases. In Figure 3 we see that if coarse-

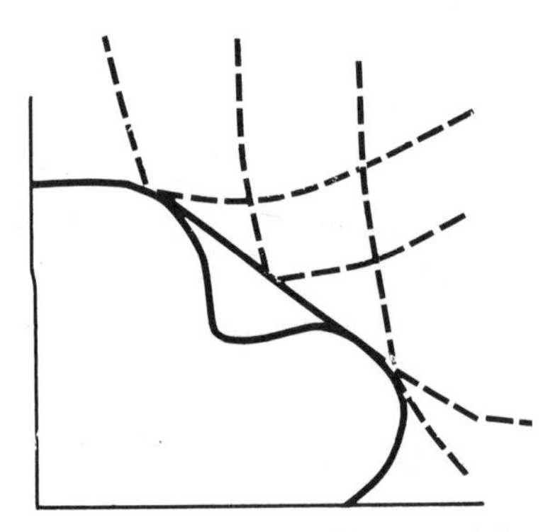

Figure 3. A pseudo-hybridization zone is created by a partially concave fitness set. Along the straight line that bounds the concavity, optimal populations are polymorphic.

grained selection occurs along a transect with changing proportions of environments, there will be a continuous cline in phenotype up to the concavity, a region of polymorphism, and then a continuous cline again. Note that if this were observed in nature it would be interpreted as a zone of secondary contact and hybridization.

6. Inbreeding flattens the genotypic fitness of Model II toward the straight line joining the homozygous points. When the fitness set is convex, this reduces fitness. But when the fitness set is concave and selection coarse-grained, fitness will be increased. This leads to an optimal level of inbreeding determined by the fitness sets of all selection components. In a flexible species with mostly convex fitness sets, the optimum inbreeding will be zero; but as flexibility decreases, the optimum inbreeding (which for our purposes would be about the same as assortive mating) increases. Thus for these species we would expect stable patterns of sympatric semi-isolates which need not be incipient species.

-6-

Some Analyses of Hidden Variability in Drosophila Populations*

JAMES F. CROW
University of Wisconsin
Madison, Wisconsin

A major aim of population genetics has been the study of genetic variability in natural populations and the ways in which this is maintained. With Mendelian inheritance, such as Drosophila has, most of the variability is not overtly expressed. There have been two principal methods of studying that part which is ordinarily hidden, and the development of both owes much to the leadership of Th. Dobzhansky. One method is the direct cytological study of chromosome polymorphisms by exploiting the convenient properties of Drosophila salivary gland chromosomes. The second is to reveal previously concealed variability by homozygosity; this can be done by inbreeding, or by crosses using marked crossover-suppressing chromosomes that achieve the same result more efficiently.

Recent developments in molecular biology have now made possible the study of genetic variability through the direct assay of proteins. This promises a totally new precision in genetic analysis of both evolutionary statics and dynamics. Exciting beginnings have been made with the discovery by Lewontin and Hubby (1966) of extensive isoallelic polymorphisms that had been suspected by some, but were hitherto incapable of detection. In the study of dynamics, it is not possible by amino acid sequence analysis to measure evolutionary rates directly in terms of gene substitutions (for reviews, see Bryson and Vogel, 1965).

I should like to report a series of studies of the earlier type. They are concerned with the rate of origin and the maintenance in the population of mutant genes, particularly those having a very small effect on viability.

*Paper number 1183 from the Genetics Laboratory. This work has been supported by the National Institutes of Health (GM07666 and GM08217).

One technique is a scheme used by Mukai (1964) whereby mutants are allowed to accumulate in a chromosome that is maintained in the heterozygous condition with as little selection as possible. The other is the standard procedure of extracting chromosomes from natural populations and making them homozygous.

The data come from several experiments by my colleagues, Drs. Terumi Mukai, Rayla Greenberg Temin, Helen U. Meyer, and Peter Dawson.

THE EFFECT OF HOMOZYGOSITY ON VIABILITY

The general mating scheme for extracting chromosomes from a wild population of *Drosophila melanogaster* and making them homozygous is shown in Figure 1. The female in generation 1 is equivalent to a fly from

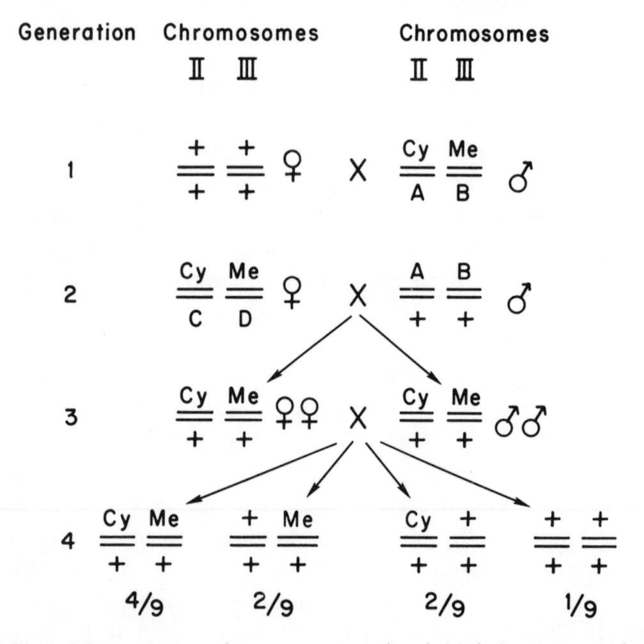

Figure 1. The mating system used to extract second and third chromosomes from natural populations and to make them homozygous. The female in generation 1 is equivalent to a fly from a natural population.

nature, being the daughter of a wild female that had already been fertilized at the time of trapping.

The flies in generation 4 are homozygous for the wild second chromosome if they have the dominant marker *Me* (*Moire* eyes), for the third chromosome if they show *Cy* (*Curly* wings), and for both second and third chromosomes if neither marker appears. The *Cy* and *Me* chromosomes are both lethal when homozygous, so the expected ratio of the four phenotypic classes *Cy Me*, + *Me*, *Cy* +, and + + is 4:2:2:1.

The symbols *A*, *B*, *C*, and *D* stand for chromosome with dominant markers. Some of these chromosomes also have translocations and carry duplications and deletions such as to eliminate by lethality or sterility most of the unwanted classes in generation 3. When the matings are carried out as shown in Figure 1, the chromosomes from nature are completely homozygous ($F = 1$) in the appropriate phenotypic classes. When the scheme is modified so that the flies mated in generation 3 are descended from two different males in generation 2, but these two males are sons of a single female in generation 1, the probability of gene identity in generation 4 is ½ (F = 0.5). Finally, if the males and females mated in generation 3 are descended from different females in generation 1, the unmarked homologues are equivalent in heterozygosity to flies in the natural population ($F = 0$). Details of the mating system used to insure that each chromosome is represented the same number of times at different F levels and a description of the special stocks used are given by Temin, Meyer, Dawson, and Crow (unpublished manuscript, 1968).

The analysis was carried out according to the model shown in Table I. The quantity S is the relative viability of the second chromosome homozygote; that is, it is the amount by which the heterozygous viability must

TABLE I

The Method of Analysis of the Effect of Homozygosity on Viability

Phenotypes	*Cy Me*	*+Me*	*Cy+*	*++*
Observed Ratios				
$F = 0$	W	X	Y	Z
$F = 1$	w	x	y	z
Expected Ratios				
$F = 0$	W	X	Y	Z
$F = 1$	W	SX	TY	$STIZ$

Estimators

$$\frac{\text{Homozygous viability}}{\text{Heterozygous viability}}: \quad S = \frac{Wx}{Xw}, \quad T = \frac{Wy}{Yw}$$

$$\text{Interaction:} \quad I = \frac{XwYz}{WxZy}$$

$$\text{Homozygous loads:} \quad s = -\ln S,\ t = -\ln T,\ i = -\ln I$$

be multiplied to give the homozygous viability. A consistent estimate of S is given by equating the observed quantities to their expectations. Thus

$$\frac{W}{X} \cdot \frac{x}{w} = \frac{W}{X} \cdot \frac{SX}{W} = S$$

and likewise for T, the third chromosome homozygous viability. By this procedure any viability effects associated with the marker chromosomes or other extraneous variables cancel out.

I is a measure of interaction. If the viability effects of second and third chromosome homozygosity are independent, the joint survival will be the product of the individual components. When there is independence, then, $I = 1$. Reinforcing epistasis or synergism is indicated by values less than 1; opposing epistasis, by I greater than 1. For a synthetic lethal, S and T are approximately 1 and $I = 0$. I can be estimated by equating observed and expected ratios, as before. In this case

$$\frac{X}{W} \cdot \frac{w}{x} \cdot \frac{Y}{Z} \cdot \frac{z}{y} = \frac{X}{W} \cdot \frac{W}{SX} \cdot \frac{Y}{Z} \cdot \frac{STIZ}{TY} = I.$$

The interactions will be discussed later.

The bottom line in Table II shows that the effect of homozygosity is to multiply the heterozygous viability by 0.67 for the second chromosome (S) and 0.66 for the third (T); that is, to reduce the viability by about one-third.

The homozygous viabilities were classified into 3 groups: mildly detrimental (viability greater than 50 per cent of that of heterozygotes), severely detrimental (10 to 50 per cent), and lethal (less than 10 per cent). The 10 per cent criterion for lethality was chosen because of the difficulty of classifying some borderline lethals with rare survivors. These "escapees" are typically weak and do not survive long after emergence, and more reproducible results are obtained by including these with the lethals. Such

TABLE II

Viabilities of homozygotes ($F = 1$) for second (S) and third chromosome (T) relative to the viability with random mating ($F = 0$). The contribution to this total from mild, severely detrimental, and lethal homozygotes is also shown. Also given are the complements, $1-S$ and $1-T$.

	S	T	$1-S$	$1-T$
Mild detrimentals (>0.5)	0.91	0.88	0.09	0.12
Severe detrimentals (0.1–0.5)	0.94	0.95	0.06	0.05
Lethals (>0.1)	0.78	0.79	0.22	0.21
Total (= product)	0.67	0.66	0.33	0.34

strains must usually have a fitness of zero in natural populations. The viability distribution of homozygotes is distinctly bimodal with very few between 10 and 60 per cent, as is shown graphically in Figure 1 of Greenberg and Crow (1960). The data in Table II are the averages of four experiments (Temin *et al.*, *op. cit.*).

Figure 2 shows graphically the manner of analysis by viability classes. When all homozygotes are counted, S or T is a measure for the second or

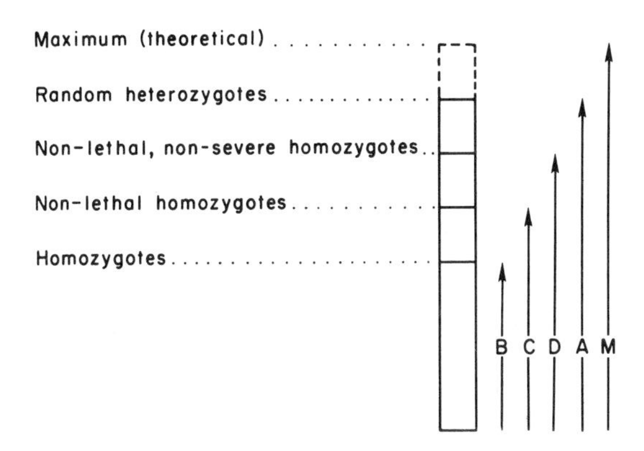

Figure 2. A diagram to illustrate the relative viabilities of homozygotes (with various viability classes eliminated), random heterozygotes, and a theoretical individual with no mutation load on the relevant chromosome.

third chromosome of B/A. (Each is measured relative to the *Cy Me* class as explained before; for example B/A is estimated for the second chromosome by Wx/Xw.) When lethals are omitted from the analysis, the same procedure gives a measure of C/A. When lethals and severe detrimentals are omitted, D/A is measured. Therefore, D/A is a measure of the effect of homozygosity for mildly detrimental chromosomes. The effect of severly deleterious chromosomes alone is C/D, estimated by $(C/A)/(D/A)$, and that of lethals alone is B/C, estimated by $(B/A)/(C/A)$. The overall homozygous viability is $(D/A)(C/D)(B/C) = B/A$, as it should be if the effects of the different classes are independent.

The results are shown in Table II. The S and T values, and their complements, are given for mild detrimentals (D/A), severe detrimentals (C/D), and lethals (B/C). There is, of course, an independent estimate of $1 - B/C$; namely, the simple frequency of nonlethal chromosomes. This will be slightly different because occasional survivors render the viability of chromosomes classified as lethal slightly greater than 0, and this has a slight influence on the B/C calculation. However, the actual frequencies

TABLE III

Homozygous loads for second and third chromosomes. The last column (s') is from an earlier experiment to show the reproducibility of the results.

	s	t	s'
Mild detrimentals, D_m	0.095	0.122	
Severe detrimentals, D_s	0.062	0.053	
Total detrimentals, $D = D_m + D_s$	0.157	0.175	0.147
Lethal, L	0.247	0.240	0.262
Total, $D + L$	0.404	0.415	0.409
D_m/L	0.39	0.51	
D/L	0.64	0.73	0.56

of lethal chromosomes are 0.230 for chromosome two and 0.225 for three, which agree very well with 0.22 and 0.21 from Table II.

The viability effects can also be expressed as homozygous loads. These are given in Table III and are computed as the negative natural logarithms of the quantities in Table II. The loads have the merit of being additive rather than multiplicative and of giving an estimate of the total effect of the genes if their independent homozygous effects were dispersed into separate individuals (Morton, Crow, and Muller, 1956; Greenberg and Crow, 1960). For example, with lethals s ($= 0.247$) is larger than $1 - S$ ($= 0.22$) since it corrects for chromosomes with more than one lethal.

The same conclusions appear as in Table II. The major viability reduction is caused by lethals. The reproducibility of results in experiments done at different times is shown by the last column which repeats earlier second chromosome data from Temin (1966).

This analysis takes into account the fact that severely detrimental chromosomes also contain some mildly detrimental genes and that the chromosome viability is thereby somewhat reduced. This reduction is included in the estimate of D_m. Likewise, the lethal chromosomes contain some mild or severe detrimentals, although these, of course, have no further effect if the chromosome is already fully lethal. Again, these detrimental effects are included where they should be, in D_m and D_s, respectively. For an explanation, see Greenberg and Crow (1960). However, an error may be introduced by misclassification if a chromosome with several detrimentals is of such low viability as to be classified as a lethal. In this case, the frequency of the lethal class may be somewhat overestimated, although this is not likely to be an appreciable error because of the small number of chromosomes in the viability range from 10 to 60 per cent. On the other hand, there is a greater possibility of underestimating the frequency of the mildly deleterious class by erroneously classifying some of them as carrying severe detrimentals.

MUTATION RATES FOR MILD DETRIMENTALS

Mukai (1964) reported that spontaneous mutations with mild viability effects occur much more frequently than lethals. The experimental procedure is shown in Figure 3. *Pm* (*Plum* eyes) and *Cy* (*Curly* wings) are

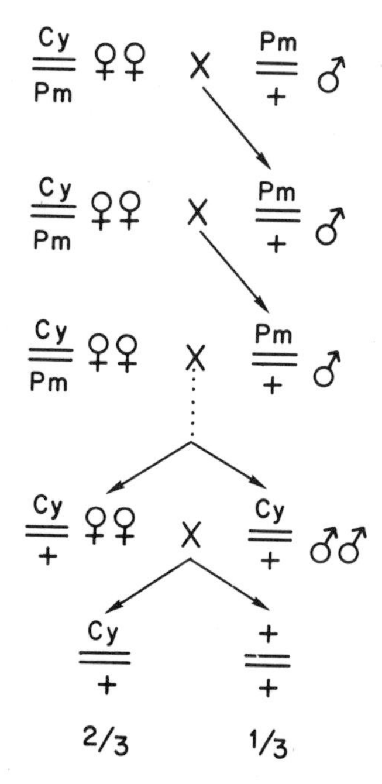

Figure 3. The mating system used to accumulate mutations in a normal second chromosome (designated by +) over many generations. Later the chromosome can be made homozygous for viability testing by the crosses shown at the bottom.

both dominant markers. The *Cy* chromosome has several crossover-suppressing inversions and is lethal when homozygous. Spontaneous mutations were allowed to accumulate for many generations with as little selection as possible during this time. At intervals of several generations, the homozygous viability effects were measured by the matings at the bottom of the figure. The experimental details are given by Mukai (1964).

In these experiments there were 104 lines. The wild-type second chromosomes in all of them were descended from a single chromosome that was tested for homozygous viability and then multiplied into the 104 lines. Any significant selection between lines is ruled out by the fact that only three were lost during the experiment. Selection within lines was minimized by growing the cultures in uncrowded conditions and by using only a single male in each generation; a second male, kept as insurance against sterility each generation, had to be used only very rarely. The total experiment involved 1.7 million flies. The procedure was replicated in a completely independent experiment completed in 1967. Table IV shows both the 1964 and 1967 results.

TABLE IV

The rate of accumulation of deleterious homozygous effects on viability in chromosomes kept heterozygous and with minimum selection. Included also are estimates of the mutation rate per chromosome for mild detrimentals (Σu) and lethals (ΣU) and the average viability decrease caused by mild detrimentals.

	1964	1967
Detrimental load/generation ($D_m = \bar{s}\Sigma u$)	0.0039	0.0035
Σu (minimum estimate)	0.16	0.14
$\bar{s}$ (maximum estimate)	0.022	0.028
Lethal mutation rate/generation (ΣU)	0.005	0.006
Detrimental load/lethal load (D_m/L)	0.78	0.58

The estimating procedures are as follows. Let s_i be the homozygous viability decrease as a result of mutation at the ith locus. Let u_i be the probability of a mutation at the ith locus during the period of mutant accumulation. The mean reduction in viability will be

$$M = \Sigma us = \bar{s}\Sigma u \quad (1)$$

where the summation is over all loci and s ($= \Sigma us/\Sigma u$) is the weighted average value of s with weights proportional to the mutation rates. The variance in viability attributable to the mutation at the ith locus is $u_i(1 - u_i)s_i^2$. Summing over all loci on the chromosome, the overall variance is

$$V = \Sigma u(1 - u)s^2 \approx \Sigma us^2 = \overline{s^2}\Sigma u \quad (2)$$

where the weighting to obtain $\bar{s}^2$ is the same as before. But $\overline{s^2} = \bar{s}^2 + v_s$, where v_s is the variance of the individual s values. Therefore, since v_s is necessarily positive,

$$V > \bar{s}^2\Sigma u \quad (3)$$

the equality being true only if the s values are constant ($v_s = 0$). From (1) and (3) we obtain the estimating equations

$$\bar{s} < V/M \quad (4)$$

and

$$\Sigma u > M^2/V. \tag{5}$$

M and V are measured directly from the mean and variance of the homozygous viabilities and Σu is converted into a rate per generation by dividing by the number of generations.

The lethal mutation rates in Table IV are very similar to those obtained by other investigators. The average for a number of studies reviewed by Crow and Temin (1964) was 0.0050. The mildly deleterious mutants are defined, as in the data of Table III, as having viability greater than 50 per cent of normal; "normal" in this case is the homozygous viability of the chromosome from which all the lines in the experiment were descended. The spontaneous rates are quite high for mildly deleterious mutations in both experiments. Whether this is because of a larger number of loci producing mild detrimentals, or whether the rate per gene is higher, or both, is not known. The estimate from equation (5) above is about 0.15 per second chromosome per generation, or roughly 0.3 to 0.4 per gamete. This is probably an underestimate for several reasons. One is that some of the mutants were probably eliminated by heterozygous selection despite attempts to keep this to a minimum. A second reason is that the assumption of constant s values is undoubtedly wrong. The estimate is likely to be particularly low if there are a large number of loci with very minute effects.

For the same reasons the estimate of $\bar{s}$, about 0.025, is likely to be too high. Presumably, the viability classes grade imperceptibly into those with effects too small to measure. The product, $\bar{s}\Sigma u$, which is the homozygous load for new mutations, is more accurately known than either of its two components.

The total mutation rate estimated here is higher than has been traditionally assumed. Partly, this is because the enormous numbers and large number of generations permit measurements of mutations with small effects. The results are not so surprising when considered in molecular terms. The opportunity for nucleotide substitutions with hardly any effect on protein function is clearly present, and a large number of mutants grading imperceptibly from clearly deleterious to those with effects too small to measure is to be expected. Completely neutral, highly multiple isoalleles would produce heterozygosity at 30 per cent of the loci with a mutation rate of $1/(10N)$ where N is the effective population number (Kimura and Crow, 1964). An amount of polymorphism more than that observed by Lewontin and Hubby (1966) could, if necessary, be explained solely by mutation of neutral genes with an effective population number of, say, 10^4 and an average mutation rate of 10^{-5}. With a mutation rate of 10^{-5} per gene, Mukai's chromosomal mutation rate requires at least 10^4 genes per second chromosome.

PARTIAL DOMINANCE OF DELETERIOUS MUTANTS

The greater dominance of mild deleterious mutants than of lethals is suggested by comparison of the D_m/L ratios in Tables III and IV. The definition of detrimentals in Table IV is the same $>50\%$ criterion used for D_m in Table III. The ratios are roughly the same in equilibrium populations as for newly occurring mutants. If the heterozygous disadvantage were proportional to the homozygous effect, the mild detrimentals would persist in the population much longer than the lethals, and the D_m/L ratio would be much larger at equilibrium than for new mutants. With constant dominance, the ratio of the equilibrium $D{:}L$ to that for new mutants is $\overline{S}/\overline{s}$, where $\overline{S}$ and $\overline{s}$ are the average selective disadvantage for lethals and detrimentals (Greenberg and Crow, 1960). Since $\overline{s}$ is estimated maximally as 0.02 or 0.03 (Table IV) and $\overline{S}$ is practically 1, the expected ratio of the $D{:}L$ ratios is 30 to 50 or more, tremendously larger than the observed value of roughly 1. Thus, as suggested earlier (Greenberg and Crow, 1960; Chung, 1962), detrimentals appear to have much greater dominance than lethals.

However, the above argument refers to loads measured from a hypothetical genotype in a theoretical population at equilibrium with recurrent mutation rate zero, but with all variability not attributable to recurrent mutation remaining the same as in the real population. The actual data are measured from chromosomes combined at random and, therefore, are equivalent to random mating in a natural population presumably somewhere near mutational equilibrium. The loads must be corrected for this discrepancy. I shall use a procedure first employed by Temin (1966).

Refer again to Figure 2. We have quantities like D/A, whereas we really want to measure quantities like D/M. But $D/M = (D/A)\,(A/M)$. Thus the corrected value for the homozygous load due to mild detrimentals is

$$
\begin{aligned}
D'_m &= -\ln(D/M) \\
&= \ln(A/D) + K \\
&= \ln(Xw/xW) + K \quad \text{(for second chromosomes with lethals and severe detrimentals omitted)} \\
&= D_m = K
\end{aligned}
$$

where $K = -\ln(A/M)$ is the mutation load expressed as viability reduction in a randomly mating population.

If we make the reasonable assumption that no appreciable contribution to the random load comes from occasional lethal homozygotes, the lethal load requires no correction, for the corrected value is then $L' = -\ln(B/C)$, the same as the uncorrected value. This is not surprising with lethals, perhaps, for a quite different reason; a lethal has 100 per cent reduction in viability, whatever the reference point.

If all the effect of the mutants were on viability (that is, if this were the sole means by which they are eliminated through natural selection), the random mating load K for independent genes would be about twice the mutation rate per gamete (Haldane, 1937). From Table IV, $2(\Sigma u + \Sigma U)$ is about 0.3, and from Table III, $D_m = 0.11$. Thus, at equilibrium the homozygous load from detrimentals $D_m + K$ is about 0.4.

The ratio R of D'_m/L in an equilibrium population to D_m/L for new mutations is

$$R = \frac{\tilde{H}\Sigma u}{\tilde{h}\Sigma U} \div \frac{\bar{s}\Sigma u}{\bar{S}\Sigma U} \quad \text{(Greenberg and Crow, 1960)} \tag{7}$$

or

$$\frac{H}{\tilde{h}} = \frac{\bar{s}}{\bar{S}} R \tag{8}$$

where the s is the selective disadvantage of a detrimental homozygote, hs is the disadvantage of a heterozygous detrimental mutant (therefore h is a measure of dominance), and u is the detrimental mutation rate. Large letters refer to corresponding quantities for lethals. The means $\tilde{H}$ and $\tilde{h}$ are harmonic rather than arithmetic. Using $D_m/L = 0.45$ for equilibrium population (Table III) and 0.68 for new mutants (Table IV), then $R = 0.45/0.68 = 0.66$. If $s = 0.025$ and $S = 1$, then by equation (8) $\tilde{h}/\tilde{H}$ is roughly 60. Correcting D_m by using $D'_m = 0.4$ reduces this to about 17. D'_m is probably over-corrected, since it assumes that all the random mating mutation load is expressed as viability reduction, whereas there is good reason to think that a substantial part may be expressed as fertility differences. Corrected or uncorrected, there is good evidence for greater partial dominance of detrimentals than of lethals. If $\tilde{H}$ is taken as 0.015, as estimated by Crow and Temin (1964), then $\tilde{h}$ is at least 0.25 by this analysis.

Wallace (1965) has questioned the conclusion that $\tilde{H}$ is positive. I shall therefore estimate $\tilde{h}$ in another way that does not involve any assumption about $\tilde{H}$. The equilibrium homozygous load for detrimentals is $\Sigma u/\tilde{h}$. Thus $\tilde{h} = \Sigma u/D'_m$. Using $\Sigma u = 0.15$ and $D'_m = 0.4$, $\tilde{h}$ is estimated as 0.37. As mentioned above, it is possible—indeed likely, in my opinion—that the major means of elimination of these mutant heterozygotes is by fertility differences rather than viability. Otherwise, it is hard to account for the fact that the survival from egg to adult under laboratory conditions not too different from those of these experiments is 90 per cent or more. Therefore, the estimate of D_m as about 0.11 is probably too small when the effects on fertility are considered. If the true value is as high as 0.7, $\tilde{h}$ is about 15 per cent.

One source of error, mentioned earlier, is the possibility of a chromosome with a large number of mild mutants being classified in the severe group. However, adding the entire severely detrimental load D_s to D_m does not change these conclusions.

Although none of these arguments is without loopholes, since each demands some untestable assumptions, I believe that together they offer strong evidence for a substantial partial dominance of mildly deleterious mutants. I should note that $\tilde{H}$ and $\tilde{h}$ are the average dominance of those mutants found in an equilibrium population, or the harmonic mean of newly occurring mutants, which would be smaller than the arithmetic mean of the dominance of new mutants (Morton *et al.*, 1956; Crow, 1964).

Further evidence for this same conclusion is offered by several reports of a positive correlation between the homozygous and heterozygous effects of the same chromosome, although it should be noted that others have failed to find the correlations. The correlation is particularly high when lethals are omitted, as would be expected if the above analysis is correct (see for example Wills, 1966).

THE AMOUNT OF EPISTASIS FOR MILD DETRIMENTALS

There is now considerable evidence for some synergistic detrimental action of homozygous chromosomes in Drosophila (for example, Dobzhansky, Spassky, and Tidwell, 1963; Spassky, Dobzhansky, and Anderson, 1965; Dobzhansky Spassky, and Anderson, 1965; Mukai, 1967). I should like to report some additional data bearing on this point.

The main population interest is in the interaction of mild detrimentals. Lethal mutants necessarily conform to a model of multiplicative fitnesses with no interaction, since the product of any number of zeros is still zero. For similar reasons, severe detrimentals will tend to show little statistical evidence of interaction on a multiplicative model. Thus, we would expect to find the drastic chromosomes showing general agreement with a model of independent multiplicative fitnesses. Furthermore, drastic genes (lethals and severe detrimentals) are individually of low frequency in the population, so that homozygosity for more than one in the same individual is a very unusual event. Therefore, interaction among drastic homozygotes will not have any appreciable influence on the properties of a randomly mating population.

The experimental design shown in Figure 1 and Table I permits an analysis of interaction between the homozygous effects of the second and third

TABLE V

The homozygous loads for mild detrimentals on the second chromosome (s), the third chromosome (t), and the interaction (i). D_m and D_m^* are by two different methods of calculation. Also shown are the homozygous viabilities, expressed in terms of the heterozygous viability for the same chromosome and the interaction I.

	s	t	i
D_m	0.095	0.122	0.027
D_m^*	0.096	0.123	0.021
	S	T	I
D_m	0.91	0.88	0.97
Spassky *et al.* (1965)	0.89	0.86	0.88

chromosomes. I shall consider only the mild detrimentals, for the reasons just given. The results are shown in Table V.

The loads s, t, and i, were computed in two ways. D_m was calculated from the total numbers for all experiments (102,000 for $F = 1$; 180,000 for $F = 0$), eliminating lethal and severely detrimental homozygous cultures and using the formulae at the bottom of Table I. These are the same calculations used for Table III. The second method of calculation, used to obtain D_m^*, was to compute such quantities as $-ln\ (Wx/Xw)$, not for the totals but for each strain separately. These values were then averaged for all strains, about 3,500 in all. Each value was weighted by the reciprocal of its theoretical variance. As can be seen, the two procedures gave almost identical results.

A positive value of i means that the viability of the double homozygote is less than the product of the two single homozygote viabilities. Thus, taken at face value, there is a slight epistasis of the reinforcing or synergistic type. However, i is not significantly different from zero in these experiments.

Table V also shows the data in another form. S and T are the same values given in Table II, and give the viability of homozygous second and third chromosomes relative to random heterozygotes. With no interaction on a multiplicative model, $I = 1$. The value, 0.97, shows a slight, nonsignificant synergism. For comparison, the corresponding data for *Drosophila pseudoobscura* (Spassky, Dobzhansky, and Anderson, 1965) are also shown. The homozygous viabilities are very close to those for *D. melanogaster*. In this case the interaction is stronger, but not very much so. Considering all the sources of error in this kind of experiment, it is likely that the two results do not differ significantly.

Whichever is nearer the true value, the synergism is not large, and its importance to the population would be slight unless many mutants were combined in the same homozygote.

Additional evidence for synergism comes from comparisons within chromosomes of the viability for $F = 1$ with that for $F = 0.5$. With independently acting genes, the homozygous load should be exactly twice as large for $F = 1$ as for $F = 0.5$. Some results are shown in Table VI. The effects of the second and third chromosomes are summed.

TABLE VI

Second chromosome (s) and third chromosome (t) combined loads for two levels of homozygosity $F = 1$ and $F = 1/2$, for mild detrimentals (D_m), severe detrimentals (D_s), and lethals (L).

	$F = 1$		$F = 1/2$
	$s + t$	$(s + t)/2$	$s + t$
D_m	0.217	0.108	0.040
D_s	0.115	0.058	0.060
L	0.487	0.244	0.239

The relevant comparison is between the last two columns. As expected, there is no evidence for interaction in the lethals and in severe detrimentals. However, there is evidence for appreciable synergism among the mild detrimentals. The value of $s + t$ for $F = 0.5$ is considerably less than $(s + t)/2$ for $F = 1$, and the difference is significant. Moreover, this is likely to be an underestimate because there was not a complete randomization of chromosomes in producing the $F = 0.5$ group. The only opportunity for crossing-over was in the wild-type female of generation 1 and, to the extent that this did not achieve randomization, the epistatic effect is under estimated.

The possibility, suggested by the data, that the interaction is greater within chromosomes than between chromosomes must await further studies to see if the effect is reproducible.

As shown in an earlier section, there is considerable evidence for partial dominance of mild detrimentals. The calculations were made as if genes act independently, but the amount of dominance so estimated is large enough that epistasis of the magnitude indicated here would not change the general conclusion. However, if there is a substantial amount of dominance, then the interaction between *heterozygous* effects becomes an important factor in the population. As has often been mentioned, this would lessen the mutation load by preferential elimination of individuals with multiple mutants (for a quantitative treatment, see Kimura and Maruyama, 1966).

Direct data on the interaction of heterozygous effects of mild detrimentals are not available, but Kitagawa (1967) has reported a significant deviation from linearity in multiple heterozygotes for lethals. Lethals are rare enough so that a Drosophila having more than three is quite unusual, so these interactions are not very important to the population. On the other hand, the apparently far greater dominance of mild detrimentals may mean that their heterozygous effects may be of the same order of magnitude as the heterozygous effects of lethals, particularly when fertility effects are conidered. In this case, the number of heterozygous mutants in the same fly may be large enough for their interactions to produce some lessening of the mutation load.

An extreme form of this kind of epistasis would produce something approximating a threshold. The data do not suggest such a strong effect, but the rather mild synergistic viability effects observed in laboratory cultures may be accentuated in a natural habitat where fertility differences become important and conditions are more rigorous. The large reduction of the mutation load under certain threshold conditions has been pointed out by Sved, Reed, and Bodmer (1967) and King (1967).

SUMMARY

In Drosophila, the depression in survival to adulthood produced by homozygosity is mainly due to recessive, or nearly recessive, lethals. The inbred load thus revealed and measured in units of lethal equivalents per chromosome is about 0.11 for mild detrimentals, 0.06 for severe detrimentals, and 0.25 for lethals.

The rate of occurrence of mildly detrimental mutants is estimated as 0.15 per second chromosome per generation. The average reduction in viability caused by these mutants when homozygous is estimated as 2.5 per cent. The nature of the necessary assumptions is such as to make the mutation rate estimate minimal and the viability reduction estimate maximal.

There is evidence showing that such "recessive" mutants are also deleterious as heterozygotes, enough so that their major effect on the population is through their heterozygous effects. The average dominance is estimated to be several times as large for mildly deleterious mutants as for lethals.

Considering data from various sources, there is evidence for some synergistic interaction of mildly deleterious mutants when homozygous. The extent of such interactions in the heterozygous effects is yet unknown, but may be large enough to cause some reduction of the mutation load.

LITERATURE CITED

Bryson, V., and H. J. Vogel, 1965, *Evolving Genes and Proteins*. Academic Press, New York.

Chung, C. S., 1962, "Relative Genetic Loads due to Lethal and Detrimental Genes in Irradiated Populations of *Drosophila melanogaster*," *Genetics*, 47:1489–1504.

Crow, J. F., 1964, "More on the Heterozygous Effects of Lethals in Drosophila Populations," *Amer. Natur.*, 98:447–449.

Crow, J. F., and Rayla G. Temin, 1964, "Evidence for the Partial Dominance of Recessive Lethal Genes in Natural Populations of Drosophila," *Amer. Natur.*, 98:21–33.

Dobzhansky, Th., B. Spassky, and W. Anderson, 1965, "Biochromosomal Synthetic Semi-lethals in *Drosophila pseudoobscura*, *Proc. Nat. Acad. Sci.*, 53:482–486.

Dobzhansky, Th., B. Spassky, and T. Tidwell, 1963, "Genetics of natural populations. XXXII. Inbreeding and the mutational and balanced genetic loads in natural populations of *Drosophila pseudoobscura*," *Genetics*, 48:361–373.

Greenberg, Rayla, and J. F. Crow, 1960, "A Comparison of the Effect of Lethal and Detrimental Chromosomes from Drosophila Populations," *Genetics*, 45:1153–1168.

Haldane, J. B. S., 1937, "The Effect of Variation on Fitness," *Amer. Natur.*, 71:337–349.

Kimura, M., and J. F. Crow, 1964, "The Number of Alleles That can be Maintained in a Finite Population," *Genetics*, 49:725–738.

Kimura, M., and T. Maruyama, 1966, "The Mutational Load With Epistatic Gene Interactions in Fitness," *Genetics*, 54:1337–1351.

King, J. L., 1967, "Continuously Distributed Factors Affecting Fitness," *Genetics*, 55:483–492.

Kitagawa, O., 1967, "Interaction in Fitness Between Lethal Genes in Heterozygous Condition," *Genetics*, 57:809–820.

Lewontin, R. C., and J. L. Hubby, 1966, "A Molecular Approach to the Study of Genic Heterozygosity in Natural Populations. II. Amount of Variation and Degree of Heterozygosity in Natural Populations of *Drosophila pseudoobscura*," *Genetics*, 54:595–609.

Morton, N. E., J. F. Crow, and H. J. Muller, 1956, "An Estimate of the Mutational Damage in Man from Data on Consanguineous Marriages," *Proc. Nat. Acad. Sci.*, 42:855–863.

Mukai, T., 1964, "The Genetic Structure of Natural Populations of *Drosophila melanogaster*. I. Spontaneous Mutation Rate of Polygenes Controlling Viability," *Genetics*, 50:1–19.

——, 1967, "Synergistic Interaction of Spontaneous Mutant Polygenes Controlling Viability in *Drosophila melanogaster*," *Genetics*, 56:579.

Spassky, B., Th. Dobzhansky, and W. W. Anderson, 1965, "Genetics of Natural Populations. XXXVI. Epistatic Interactions of the Components of the Genetic Load in *Drosophila pseudoobscura*," *Genetics*, 52:653–664.

Sved, J. A., T. E. Reed, and W. F. Bodmer, 1967, "The Number of Balanced Polymorphisms That Can be Maintained in a Natural Population," *Genetics*, 55:469–481.

Temin, Rayla G., 1966, "Homozygous Viability and Fertility Loads in *Drosophila melanogaster*," *Genetics*, 53:27–46.

Wallace, B., 1965, "The Viability Effects of Sponteneous Mutations in *Drosophila melanogaster*," *Amer. Natur.*, 99:335–348.

Wills, C., 1966, "The Mutational Load in Two Natural Populations of *Drosophila pseudoobscura*," *Genetics*, 53:281–294.

-7-

Polymorphism, Population Size, and Genetic Load*

BRUCE WALLACE
Cornell University
Ithaca, New York

INTRODUCTION

The purpose of this paper is to examine one of the ironies of genetic load theory. Calculations can lead to the apparent conclusion that a given segregational or balanced load is too large for a population to bear; identical calculations are unable to lead to any definite conclusion concerning the ability of a monomorphic population to bear what may appear to be a much larger load. Thus, Kimura and Crow (1964) have shown that a population might maintain eight alleles at a locus if each homozygote had a fitness of 0.99 relative to a fitness of 1.00 for heterozygotes. If a population were to maintain a multiple allelic series of this sort at each of five thousand gene loci, its genetic load would be 0.998. An average survival of only 0.002 appears to be too small to permit the continued existence of a *Drosophila* population and, consequently, Kimura and Crow suggest that these flies are homozygous for a majority of genes.

Observations of Lewontin and Hubby (1966) on the gene-enzyme polymorphisms of *D. pseudoobscura* have reopened the question of segregational loads once more. Sved, *et al.* (1967), King (1967), and Milkman (1967) have, each in his own way, taken exception to the procedure used by Kimura and Crow in calculating genetic load. The irony we wish to point out here, however, is that the homozygous flies envisaged by Kimura and Crow need not be homozygous for alleles other than those postulated in their model. Now, a population homozygous at five thousand loci for alleles that confer a relative fitness of only 0.99 on their carriers would seem to have a mean survival of only 10^{-22}, an average much smaller than the 2×10^{-3} cited earlier. Nevertheless, load theory is unable to say

*This paper was prepared while the author's research was supported under contract No. AT-(30-1)-2139. U.S. Atomic Energy Commission.

whether this calculated load does exist, or whether or not the homozygous population would survive.

In the following sections I will attempt to relate autoregulation of populations by density-dependent factors (Haldane, 1953; Nicholson, 1955) to what I call "hard" and "soft" selection pressures. Furthermore, I will argue that the existence of density-dependent factors enables a population to meet increased demands of selection by a contraction in population size as well as by an increase in numbers of progeny. I will attempt to show that the apparent ability to assign causes—genetic as well as environmental—to the deaths of zygotes in a population is largely an artifact of static concepts. Because it is a static concept, I will argue that genetic load cannot be used to predict the extinction of a population. Extinction is a dynamic process.

THE USUAL MODEL OF A BALANCED POLYMORPHISM

A simple numerical model illustrating a balanced polymorphism resulting from a superior fitness of heterozygous individuals is illustrated in Figure 1. The model is based on two alleles, *A* and *a*. The relative adaptive values of *AA, Aa,* and *aa* individuals are 0.8, 1.0, and 0.4, respectively. If the frequencies of the two alleles in the population are p and q, these frequencies will change each generation until equilibrium frequencies, p and q, of 0.75 and 0.25 have been established. Thus, a balanced polymorphism consisting of some 56% *AA*, 38% *Aa*, and 6% *aa* individuals will be perpetuated in the population.

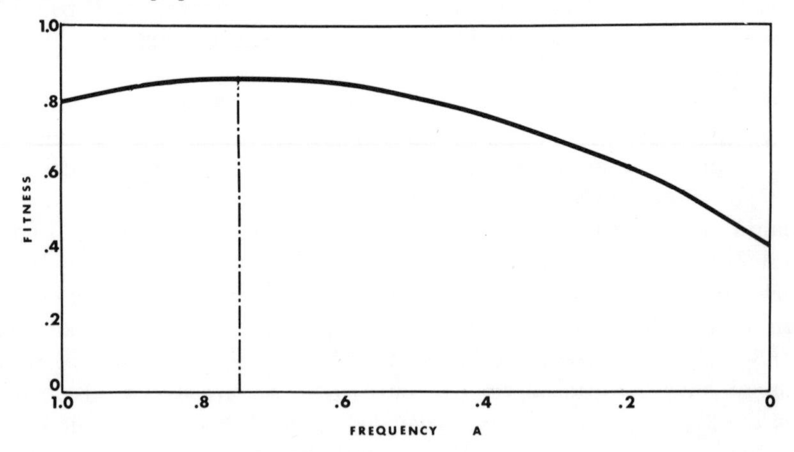

Figure 1. The average fitness of a population as a function of gene frequency when the adaptive values of *AA, Aa,* and *aa* individuals are 0.8, 1.0, and 0.4.

The polymorphism established by natural selection imposes upon the population a genetic load (more precisely, a "segregational" or "balanced" load, one of a large number of possible genetic loads). The genetic load of a population has been defined (Crow, 1958) as "the proportion by which the population fitness . . . is decreased in comparison with an optimum genotype." The average fitness of a population is calculated as the sum of the products of frequencies of individual fitnesses times the fitnesses themselves. For frequencies of *A* equal to 100 per cent or 0 per cent, the fitness of the population in our example equals 0.8 and 0.4. For an intermediate (non-equilibrium) frequency such as 50 per cent, the population fitness equals (0.25) (0.8) + (0.50) (1.0) + (0.25) (0.4) or 0.80. Finally, the maximum fitness of the population can be shown to coincide with the equilibrium gene frequencies; in the example shown in Figure 1, the maximum fitness equals 0.85.

The average fitness of a population is less than that of the optimum genotype. The genetic load imposed upon the population by this polymorphism is, by the definition given above, 0.15. The presence of the selectively inferior homozygotes, *AA* and *aa*, in the population results in the death of 15 per cent of all zygotes above and beyond the death that would occur if the population consisted entirely of *Aa* individuals or, better, of some equally fit homozygote such as A′A′.

The genetic load consisting of the excessive mortality of *AA* and *aa* homozygotes for the *A*-locus alone in our sample is 15 per cent. At how many other loci could similar loads be tolerated simultaneously? If the one locus permits the survival of only 85 per cent of those zygotes which would otherwise survive, two loci would permit the survival of only 72 per cent; three, 61 per cent; ten, 20 percent; twenty, 4 percent; forty, 0.2 per cent; and fifty, 0.03 per cent. Since *Drosophila* females lay hundreds (not thousands) of eggs, at best, the maximum number of polymorphisms of the sort we have described which could be borne at one time would lie between 20 (four eggs surviving per hundred) and 40 (two eggs surviving per thousand). Now, 40 loci is a minute fraction of the total number of approximately 10,000 gene loci of the *Drosophila* genotype.

"HARD" AND "SOFT" SELECTION

The calculations that lead to the above estimation of genetic load are based on what I call "hard" selection. Lethals, for example, impose a form of hard selection on individuals. A balanced lethal system kills one-half of all zygotes; in the absence of sophisticated remedial treatment, these individuals cannot be saved. Two independent balanced lethal systems would kill three-fourths of all zygotes; ten would kill all but one in every 1,024 zygotes. Man surely could not maintain a balanced lethal system on each of ten different chromosomes.

The impact of hard selection on a population is shown on the right in Figure 2. Only that portion of a population extending to the right of the vertical line survives and reproduces; the cross-hatched portions of the curves represent dead or non-reproducing individuals. The vertical line exemplifying hard selection has been drawn in a constant position so that the mean fitness of the populations varies in respect to it. Thus, a population containing ten independent balanced lethal systems would lie far to the left, with only a small portion of all individuals surviving. Conversely, a population containing no lethals would lie far to the right with no in-

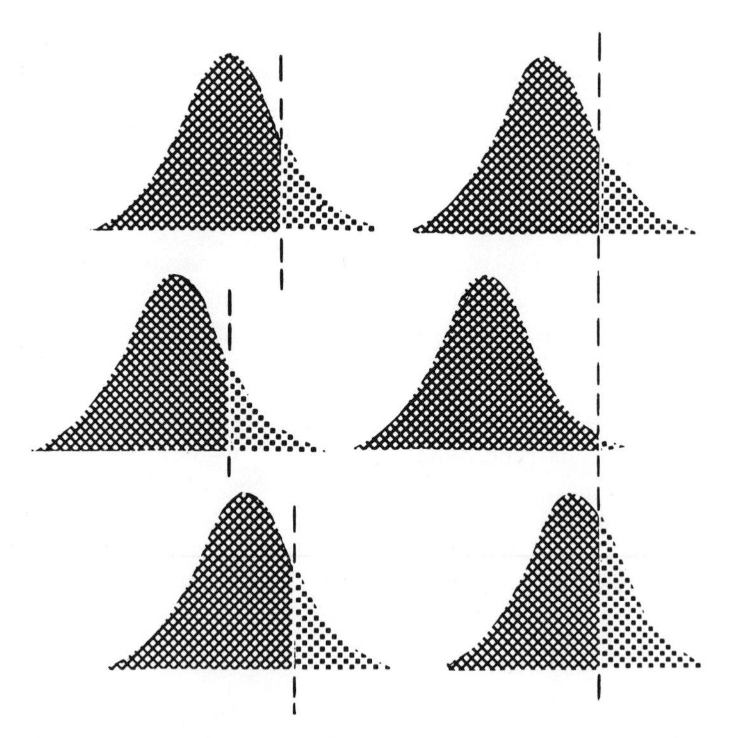

Figure 2. An attempt to illustrate "hard" and "soft" selection. Stippled portions of each curve represent individuals surviving until parenthood; cross hatched portions represent those eliminated by premature death or infertility. The three distribution curves on the right represent populations that vary in relation to a constant, unyielding ("hard") selective agent; a population that lies too far to the left may leave few if any surviving individuals. The distribution curves on the left illustrate selective forces that permit a relatively constant proportion of individuals to survive despite the precise position of the distribution curve on some arbitrary fitness scale.

dividuals dying. The converse as stated is ridiculous and therefore serves to introduce the notion of "soft" selection. As long as the number of zygotes in an equilibrium population exceeds the number of their parents, a certain fraction of them must be eliminated or rendered sterile before maturity whether lethals are present or not.

The curves at the left in Figure 2 represent "soft" selection. According to these curves, the surviving and reproducing portion of a population is very nearly constant despite the position of the population on some arbitrary fitness scale. Selection, as the shifting vertical lines suggest, yields with the nature of the population so that individuals in the upper tail of the fitness distribution of each population do, in fact, become the parents of the next generation.

An illustration of soft selection can be cited from unpublished data on the numbers of pupae formed in replicate vial cultures of *D. melanogaster* (Figure 3). The culture medium used in this experiment consisted of agar and brewer's yeast; anywhere from 10 per cent to 30 per cent (depending upon the strain of flies used) of all pupae formed in these vials eventually drowned in the soft medium. A chi-square test of homogeniety among replicate vials of the total number of pupae (drowned + undrowned) per vial revealed that the vials were heterogeneous (chi-square = 27.32; 14 df; p = 0.02). In contrast, a similar test of the numbers of undrowned pupae in the same vials revealed that these numbers were more uniform than one would expect by chance (chi-square = 5.89; 14 df; p = 0.97; the probability of obtaining results as uniform or more uniform by chance is, consequently, 0.03). Numbers of pupae *more diverse* than those expected by chance were reduced through ecological conditions in the vials to numbers *more uniform* than those expected by chance. Despite varying initial numbers, a constant number of pupae tended to survive and give rise to adult flies in these vials. If this constant number equals a fraction x of the zygotes initially present, then the genetic load for these populations could not exceed $1-x$; calculations suggesting a greater load would be meaningless if applied to the populations in these cultures.

A numerical example can be used to illustrate the point made immediately above. Suppose a certain test reveals that the fitnesses of individuals of types A and B can be represented as 1:0.001. Suppose, too, that many vials are set up in each of which there are one larva of type A and 100 larvae of type B. The genetic load of these populations can be calculated as $1 - [(0.01 \times 1) + (0.99 \times 0.001)]$ or 0.989. Now, suppose that 26 larvae can survive to maturity in each vial. If one of these is the type A larva, the other 25 will be type B. Consequently, the fitnesses of the two types are really as 1:0.25 and the genetic load is only 0.7425. Genetic load, like the phenotype of an individual, is determined not by genotypes alone but by these in relation to the environment under which they are found.

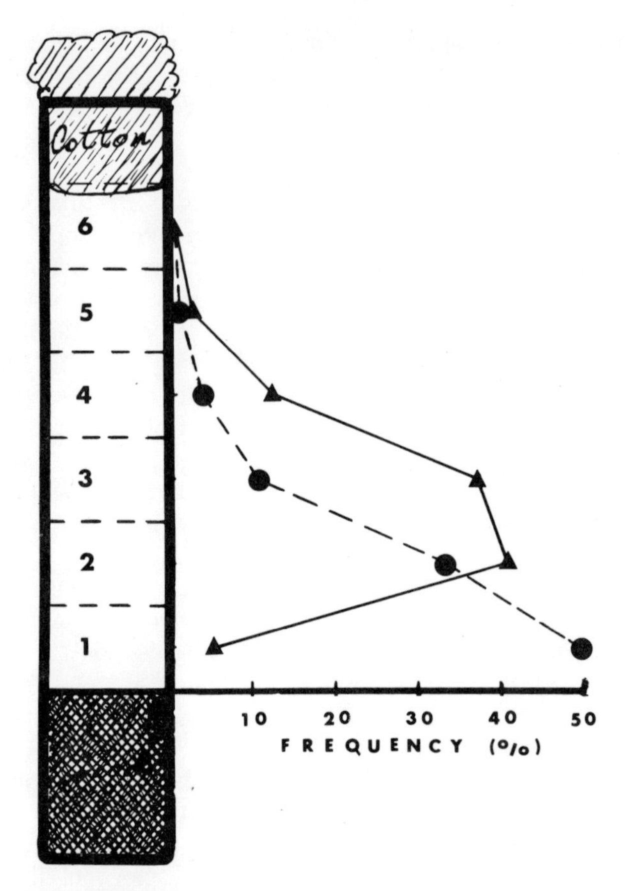

Figure 3. The distribution of pupation sites within a Drosophila culture. Triangles: Distribution in five vials yielding $bw^{D}/+$ individuals. Circles: Distribution in a sixth vial contaminated with wild-type flies of unknown origin.

POPULATION GROWTH AND DARWINIAN FITNESS

No population can continue to grow in numbers indefinitely. Calculations purporting to show that the progeny of a single pair of house flies can within one summer cover the entire earth to a depth of several meters are meaningless. Female flies do indeed lay several hundred eggs, but a large proportion of these hatch and reach maturity only if the number of flies in a neighborhood happens to be low. The proportion of successfully surviving individuals becomes less and less as the population increases in size. If we neglect males, we can say that the population grows in size until the number of adult daughters produced in one generation equals that of the mothers that produced them. When the ratio of daughters to mothers equals 1.00, the population has reached a constant size. Figure 4 illus-

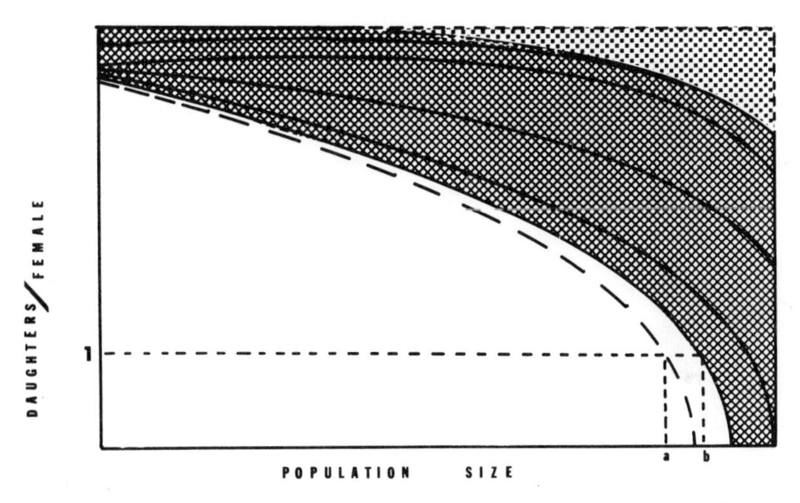

Figure 4. The ratio of surviving adult daughters to mothers as a function of population size. When the ratio equals 1.00, the population ceases to increase (b). A novel source of additional premature mortality does not necessarily eradicate an equilibrium population; the excess mortality can be met by a contraction in population size (a).

trates the relation between the ratio of daughters to mothers and population size; it is based upon notions expressed by Haldane (1953) and Nicholson (1955). The unshaded portion represents daughters surviving to adulthood; the cross-hatched portion represents individuals that die young or fail to reproduce; the stippled portion suggests that the actual number of eggs produced per female may decline as the population grows in size. (By "size" I mean something related to numbers of individuals; whether this something is actual number, density, or some other variable related to number, is unknown.)

The relationship illustrated in Figure 4 is an interesting one. It promotes a *stable* equilibrium size for populations. If a population is smaller than size *b*, for example, it will grow because the number of daughters left per mother is greater than 1.00. If, on the other hand, the population is larger than *b*, it will shrink because the number of daughters per mother is less than 1.00 in this case.

A second interesting feature of a population maintained according to the scheme shown in Figure 4 is that an environmental alteration that increases the mortality of daughters does not necessarily cause the extinction of the population. Unless the change is an abrupt or severe one, the population merely shrinks in size (*a* in Figure 4) until once more the ratio of daughters to mothers is 1.00. In effect, one cause of mortality has been swapped for another; the means by which the swap is made is a change in population size.

Relative Darwinian fitness has been defined (Dobzhansky, 1967) as "the average contribution which the carriers of a genotype, or of a class of genotypes, make to the gene pool of the following generation relative to the contributions of other genotypes." Adaptive values are measures of relative Darwinian fitnesses. Thus, in our earlier example, the adaptive values of *AA, Aa,* and *aa* individuals were given as 0.8, 1.0, and 0.4. If we neglect males, we can interpret these values in the following way: for every fertile, adult daughter left by an *Aa* mother, an *AA* mother leaves an average of 0.8 daughters while an *aa* mother leaves an average of only 0.4 daughters.

Figure 4 shows that, at equilibrium, each reproducing female in a population leaves an average of one daughter. Thus we can say that at equilibrium the average Darwinian fitness of a population is 1.00. The average fitness of the population shown in Figure 1 was said to be 0.85 when the frequencies of *A* and *a* had reached their equilibrium values of 0.75 and 0.25. This average is based on the traditional assignment of an adaptive value of 1.00 to the genotype with maximum fitness. The absolute Darwinian fitnesses of genotypes *AA, Aa,* and *aa* in a population at equilibrium in respect both to population size and gene frequencies must be 0.94, 1.18, and 0.47. At these values, the population will remain at a constant size, since the sum of the products of the frequencies of the three genotypes times their fitnesses (Darwinian fitnesses or adaptive values) equals 1.00; the average number of daughters per mother needed to maintain a constant population size. The relative Darwinian fitnesses or adaptive values of the three genotypes are unchanged.

The argument presented in the preceding paragraph says nothing about the approach of a population to its equilibrium size because, during periods of change, the average Darwinian fitness of a population is not 1.00. In the case of the population containing genotypes *AA*, *Aa,* and *aa* with equilibrium fitnesses of 0.94, 1.18, and 0.47, for example, we do not know whether these average numbers of daughters per female arise from averages of 200, 150, and 75 eggs per female or from averages such as 100, 250, and 400. During times of little mortality when nearly all eggs survive, these two arrays of fecundities would lead to markedly different patterns of gene frequency changes. Equilibrium gene frequencies, however, are determined by adaptive values under equilibrium conditions.

MONOMORPHIC AND POLYMORPHIC POPULATIONS: A HYPOTHETICAL EXAMPLE

In the preceding section we recognized that the mean Darwinian fitness of an equilibrium population must be 1.00. We recognized, too, that this mean value is the weighted average of the relative adaptive values of various genotypes. And so we can now discuss a matter that could be handled only awkwardly by the model illustrated in Figure 1; namely, the existence and fitnesses of populations consisting entirely of *AA* or *aa* individuals. According to the earlier model, these populations would have fitnesses of 0.8 or 0.4 as the figure shows. Alternatively, since at these gene frequencies (100% or 0% *A*) only one genotype is present in a population, a monomorphic population would automatically assume, under present load calculations, an average fitness of 1.00. Indeed, the information upon which Figure 1 is based is unable to tell us whether either of the monomorphic populations or the polymorphic one can sustain itself; the calculations used in constructing the parabola in the figure may, for all we know, be part of a meaningless numerical game.

The following hypothetical example illustrates the advantage gained by expressing the average fitness of a population (rather than that of the optimal genotype) as 1.00, and interpreting this value as the mean number of daughters per mother in a population of constant size. In Figure 5 are a

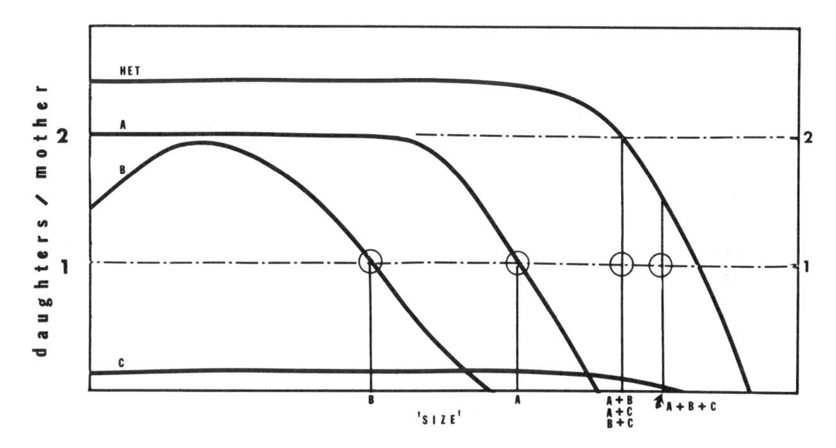

Figure 5. The ratio of daughters to mothers as a function of population size in several hypothetical monomorphic and polymorphic populations.

series of curves representing the relation between fitness (measured as daughters per mother) and population "size." The curves labeled *A*, *B*, and *C* represent individuals homozygous for three different alleles; the one labeled *HET* represents their various heterozygous combinations. The fitness scale has been chosen in an arbitrary manner suitable for illustrating the points we wish to make in the following discussion.

The points brought out by Figure 5 are the following: Populations monomorphic for alleles *A* and *B* could perpetuate themselves successfully but would do so at two quite different equilibrium sizes; a population monomorphic for *C* would not be self-sustaining under the circumstances shown in the figure. If a second allele were introduced into a population that was monomorphic for either *A* or *B*, the population size would increase. As the size of the now polymorphic population increased well beyond the equilibrium size for *A* (and, hence, for that of *B*, also), all homozygous individuals would behave as effective lethals (zero fitness). In this case, population size would become stable when heterozygous females produced an average of two daughters each, for then the mean fitness of the population would be 1.00. If all three alleles were present in a single population, the size would increase until the fitness of each heterozygous female dropped to 1.50; since one-third of the population would be homozygous for *A*, *B*, or *C* (zero fitnesses), the average fitness of the population would then be 1.00.

The above example emphasizes two additional points. An analysis of equilibrium populations polymorphic for the combinations *A* and *B*, *A* and *C*, or *B* and *C*, would reveal that each is effectively a balanced lethal system; consequently, one would infer that populations monomorphic for *A*, *B*, or *C*, could not exist. This inference would be true only for allele *C*.

The second point is the completely arbitrary manner in which genetic load varies among the populations described in Figure 5. Populations monomorphic for *A*, *B*, or *C*, would have no genetic load in the formal sense; only two of these populations could maintain themselves, however. The two that could exist would do so at different equilibrium sizes. The paired combinations of alleles would give rise to larger populations than either of the monomorphic ones; in the process they would acquire a genetic load of 0.50, a load not found in either of the smaller monomorphic populations. Finally, the three alleles together would give rise to a still larger population but one with a somewhat lower genetic load (0.33). Thus, in comparison with a smaller population, the larger one may have the same, a smaller, or a larger genetic load.

A final point that can be made explicit at this time is the relation between Figures 4 and 5. The curve drawn in Figure 4 represents the mean response of fitness to population size for an entire population. There is no reason, however, why every genotype within the population should follow precisely the pattern that describes the mean. The curves in Figure 5 represent a dissection of that in Figure 4. They show that the fitnesses of individual genotypes can react in a variety of ways to changes in population size. Each of these reactions modifies the relative fitnesses or adaptive values of the genotypes, their relative frequencies at equilibrium, and the overall "size" of the population at equilibrium as well.

MONOMORPHIC AND POLYMORPHIC POPULATIONS: LABORATORY POPULATIONS OF *D. PSEUDOOBSCURA*

During the past quarter of a century, Dobzhansky and his colleagues have made phenomenal studies on the relative fitnesses of inversion types in populations of *D. pseudoobscura* (Dobzhansky 1947, 1948, and 1950). The details of the experimental results vary according to the gene arrangements studied, their geographic origins, and whether the laboratory populations contain inversions from single geographic localities or interlocality mixtures. Almost without exception, studies of laboratory populations containing two or more different gene arrangements of chromosome 3 from a single geographic locality reveal that inversion heterozygotes are superior in fitness to homozygotes. Table I lists the results of studies of gene ar-

TABLE I

The relative adaptive values of various genotypes in experimental populations of *D. pseudoobscura* monomorphic or polymorphic for the ST, AR, and CH gene arrangements from Piñon Flats, California. (after Wright and Dobzhansky, 1946)

Gene arrangements in population	Genotype						
	ST/ST	ST/AR	ST/CH	AR/AR	AR/CH	CH/CH	$\overline{W}$
ST	1.000	–	–	–	–	–	1.000
AR	–	–	–	1.00	–	–	1.000
CH	–	–	–	–	–	1.00	1.000
ST + AR	0.81	1.00	–	0.50	–	–	0.862
adjusted	0.94	1.16	–	0.58	–	–	1.000
ST + CH	0.85	–	1.00	–	–	0.58	0.889
adjusted	0.96	–	1.12	–	–	0.65	1.000
AR + CH	–	–	–	0.86	1.00	0.48	0.890
adjusted	–	–	–	0.97	1.12	0.54	1.000
ST + AR + CH	0.43	1.30	1.00	0.05	0.71	0.21	0.799
adjusted	0.54	1.63	1.25	0.06	0.89	0.26	1.000

rangements from Piñon Flats, California (Wright and Dobzhansky, 1946). Laboratory populations monomorphic for each of the three gene arrangements—Standard (ST), Arrowhead (AR), and Chiricahua (CH)—can be maintained without difficulty and, consequently, have been included in the table. In addition to the estimates of relative fitness based on a maximum value of 1.00 which is assigned to the optimal genotype, we have adjusted these relative fitnesses in the table so that the mean fitness of each of the seven possible monomorphic and polymorphic populations equals 1.000.

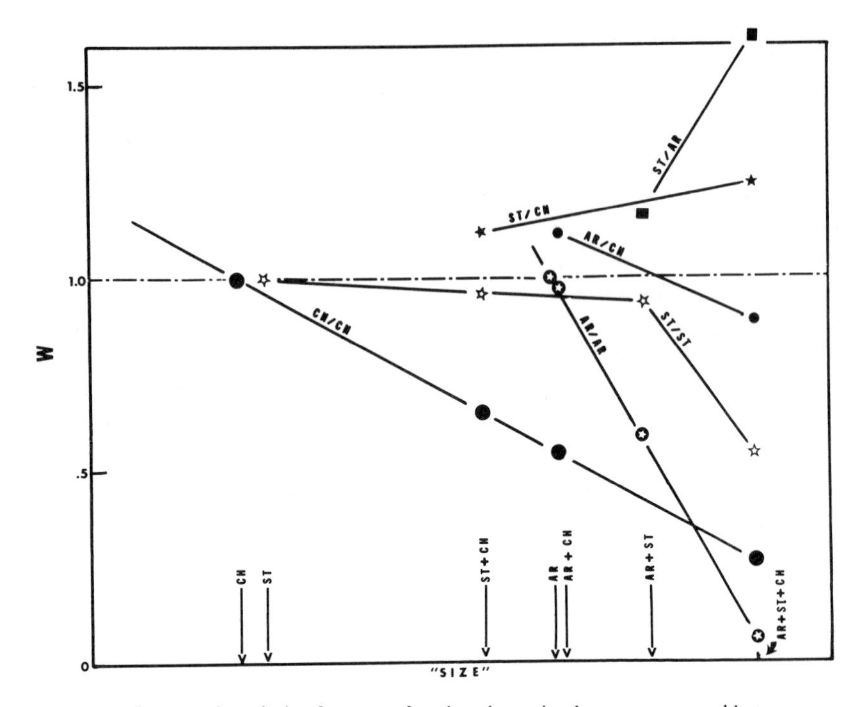

Figure 6. Data on the relative fitnesses of various inversion homozygotes and heterozygotes in laboratory populations of *D. pseudoobscura* used to reconstruct their relation to population size (see text).

The immediate problem is to use the information in Table I to construct a figure analagous to Figure 5. This has been done in Figure 6. The base line for Figure 6 was provided by CH/CH. The leftmost large solid circle lies on $W = 1.00$ and represents the monomorphic CH population. The rightmost large solid circle lies on $W = 0.26$ and represents the fitness of CH/CH individuals in the 3-arrangement polymorphic population. These two points are connected by a straight line. The two places where this line intersects fitnesses of 0.54 and 0.65 locate the sizes assigned to AR–CH and ST–CH polymorphic populations.

Two points are now known for the fitnesses of AR/AR individuals: 0.06 in the 3-arrangement polymorphic population on the far right and 0.97 in the AR–CH population. A straight line through these points intersects W = 1.00 and, hence, locates the size of monomorphic AR populations. Furthermore, the intersection of this line and that representing fitness of 0.58 locates the AR–ST polymorphic population. At this time, all points except that representing the monomorphic ST population are fixed; the size of the monomorphic ST population is located by extending the line through the values of ST/ST observed in ST–CH and ST–AR polymorphic populations.

The procedure described above is admittedly primitive. Nevertheless, it represents the best that can be done with the data at present. Some of the fitness estimates (especially those of the six genotypes in the 3-arrangement polymorphic population) are quite unreliable. The collection of curves shown in Figure 6, however, represents an approximation to the hypothetical curves of Figure 5; both represent an attempt to reveal the fitness curves of individual genotypes that make up the generalized curve of Figure 4.

In constructing the curves shown in Figure 6, CH/CH was used to establish an initial base line, AR/AR was then used to construct a second one, and finally a curve for ST/ST was drawn. The last curve was fixed by decisions made in constructing the first two. The order (CH/CH—AR/AR—ST/ST) in which the homozygotes were chosen to prepare Figure 6 was arbitrary; there are six possible orders, all of which would seem to be equally valid for the task. Is there, in fact, any basis for deciding which order of analysis is the correct one to use?

An attempt to identify the correct reconstruction of the fitness curves for the three homozygous genotypes of *D. pseudoobscura* is shown in Figure 7. The most profitable order in which to choose homozygotes is that which results in a smooth rather than an irregular curve for the third homozygote. Furthermore, the curves generally should run from upper left to lower right in the case of homozygotes. The lowest fitnesses observed for homozygous genotypes were found in the 3-arrangement polymorphic population; to maintain smooth curves, this population must be located at the far right in the diagrams. Neither homozygote can have a fitness exceeding 1.00 in a 2-arrangement polymorphic population at equilibrium; thus, the monomorphic populations with their fitnesses of 1.00 must fall at the left-hand ends of the various curves.

Of the six diagrams shown in Figure 7, only two (CH/CH—AR/AR—ST/ST and AR/AR—CH/CH—ST/ST) give smooth curves sloping generally from upper left to lower right. These two diagrams are virtually identical in their details. In the remaining four diagrams, the last homozygous genotype plotted forms a peaked line; the insertion on the left of fitness 1.00 for the monomorphic population would result in a definite zigzag line of doubtful meaning. Consequently, it seems appropriate to regard the curves shown in Figure 6 as meaningful, but not necessarily precise, representations of fitness-population size relationships in laboratory populations of *D. pseudoobscura*. Obviously, the curves representing the three homozygotes are relative so that the choice of a concave curve for either

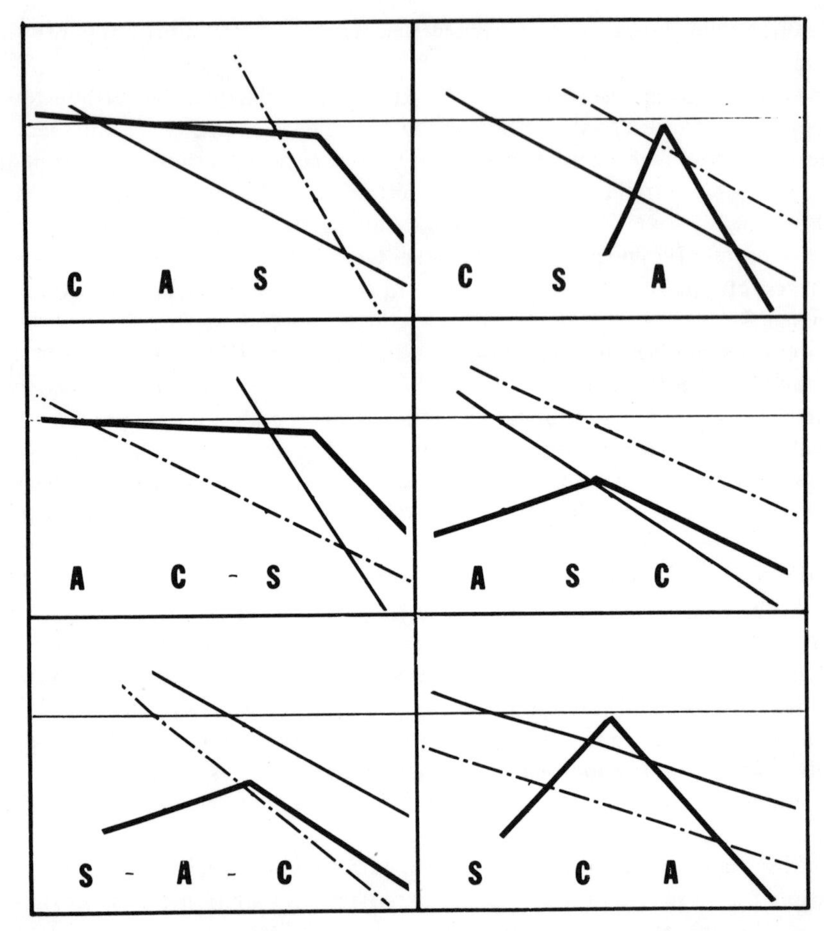

Figure 7. Construction of curves comparable to those of inversion homozygotes shown in Figure 6. C = Chiricahua, A = Arrowhead, and S = Standard (see text for explanation).

CH/CH or AR/AR, or both, would have resulted in a curve for ST/ST that would have been more nearly linear than that shown in Figure 6.

If the mean fitness of a 2-arrangement polymorphic population equals 1.00 at equilibrium, neither homozygote can possess a fitness greater than 1.00. Our insistence, then, that the curves representing the fitnesses of homozygous individuals run from upper left to lower right in Figures 6 and 7 causes polymorphic populations to exceed the corresponding monomorphic ones in size. This forced relationship seems to agree with empirical observations; Beardmore, *et al.* (1960) have reported that polymorphic populations are generally larger than monomorphic ones. The curves in Figure 6 reveal, too, that the near-zero fitness of AR/AR individuals in the

3-arrangement polymorphic population in no way precludes the existence of monomorphic AR/AR populations; in fact, according to the figure these populations appear to be somewhat larger than either monomorphic ST or monomorphic CH populations.

An attempt has been made in Figure 8 to determine the effect on populations of *D. pseudoobscura* of a general relaxation of environmental conditions. This has been accomplished by superimposing the curves representing the fitness population size relationships of the various homozygotes and heterozygotes on a new fitness scale; the line representing a mean Darwinian fitness of 1.00 has been lowered somewhat from its original position. That this does reflect less stringent environmental conditions is attested by the resultant general increase in population sizes. The monomorphic populations establish new equilibria where the appropriate curves intersect the line representing a mean fitness of 1.00; the polymorphic ones where new average fitnesses equal 1.00. (The figure as it has been drawn shows AR/ST rising rapidly to the right. Since the fitness of these flies appears to exceed 2.00, it has been impossible to show a new equilibrium size for AR–ST or AR–ST–CH polymorphic populations. It does not seem

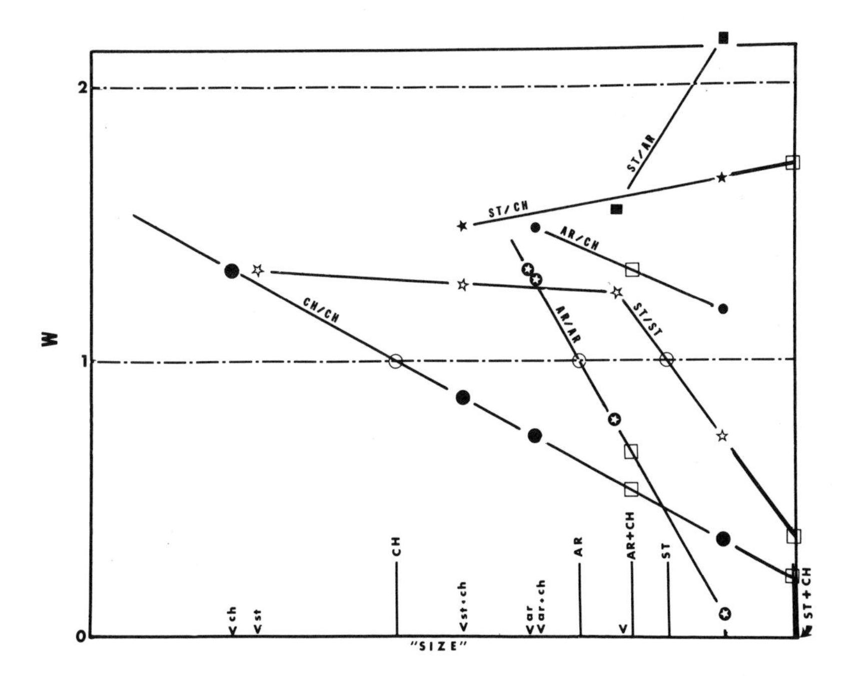

Figure 8. An attempt to determine the effect of relaxed selection on populations whose fitness-population size relationships are those shown in Figure 6. "Relaxed selection" has been achieved by altering the fitness scale.

strange that an anomaly of this sort should be encountered in the crude analysis of the sort we are making.)

The postulated easing of environmental conditions results in a general increase in population size, an increase revealed in Figure 8 by a comparison of the positions of corresponding small and large letters. Furthermore, it has resulted in more similar fitnesses of CH/CH individuals and AR/AR and ST/ST ones in the two polymorphic populations, CH–AR and CH–ST. Thus, an improvement in environmental conditions would tend to change the frequency of CH from earlier low equilibrium frequencies in these two populations (21 per cent and 26 per cent according to the data listed in Table I) to new ones of nearly 50 per cent. By coincidence or otherwise, this is very nearly the sequence of events that occurs at Piñon Flats during the spring and early summer months; the cyclic changes in the frequency of CH in the population of *D. pseudoobscura* can be generated by raising the lowering the line representing W = 1.00 as in Figures 6 and 8. Changes in relative fitnesses resulting from shifting the level at which W = 1.00 also agree with an observation by Birch (1955) that the relative fitness of CH/CH individuals in polymorphic populations is improved if larval overcrowding is avoided. Ironically, the relative fitnesses that are encountered at equilibrium after the improvement of the environment illustrated in Figure 8 are such that the genetic loads of both AR–CH and ST–CH polymorphic populations have increased. Thus, in our analysis of *D. pseudoobscura*, just as in our earlier hypothetical example, we find that the size of the genetic load (as defined mathematically) has little bearing on the state of affairs within populations. Selection operates to stabilize the average fitness (measured in daughters per mother) of a population at 1.00; the value taken on by genetic load as a result of this selection is largely immaterial. Some zygotes—all but two per parental pair in sexually reproducing species—must be eliminated or rendered sterile during the pre-adult stages of life; the cause assigned to this elimination by an outside observer following the stabilization of population size is of little consequence to the population itself.

GENE FREQUENCY-FITNESS RELATIONSHIPS

The generally accepted relationship between gene frequency and the average fitness of a population in the case of a two-allele balanced polymorphism has been illustrated in Figure 1. In the meantime, we have learned (1) that Figure 1 illustrates a density-independent relationship, and (2) that many polymorphisms are markedly density-dependent. The density-dependence of polymorphisms makes the simple parabolic relationship between fitness and gene frequency inadequate; population size as well as gene frequency must be specified for non-equilibrium conditions in the case of density-dependent polymorphisms.

An attempt to illustrate the approach of a density-dependent polymorphism to its equilibrium has been made in Figure 9. On the left,

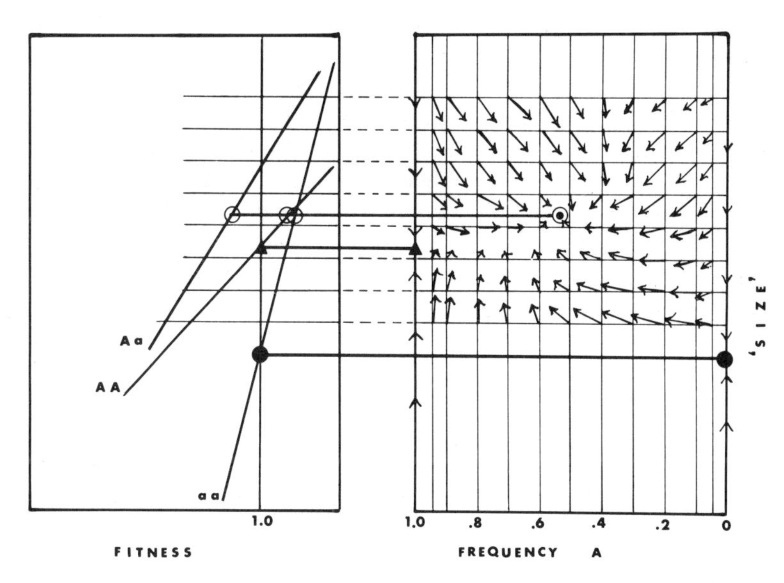

Figure 9. Two charts that relate fitness to gene frequency through a common axis, population size; fitness must be related to gene frequency in this way in a freely breeding, equilibrium population.

tipped on end, is a chart illustrating the fitnesses of three genotypes—*AA*, *Aa*, and *aa*—in relation to population size. The lines (drawn straight for simplicity) shown in the chart are not unlike those of AR/AR, AR/CH, and CH/CH of Figure 6. There are three possible equilibrium sizes: (1) that of a population monomorphic for *a*; (2) that of a population monomorphic for *A*; and (3) that of a population polymorphic for *A* and *a*. The relative fitnesses of the three genotypes in the polymorphic population enable one to calculate equilibrium gene frequencies for this population.

The chart on the right represents gene frequency and population size. Three equilibrium frequencies and sizes are known from the left-hand chart; these are the two monomorphic populations (100% and 0% A) and the polymorphic population whose equilibrium frequencies of *A* and *a* can be calculated (about 53% A). Now, non-equilibrium populations, since they represent temporary or artificial situations, can be constructed on the basis of any combination of size and frequency. How does each of these unstable populations evolve? Simple calculations on a desk calculator give the results indicated by arrows in the right-hand chart. There is no simple parabola leading to the equilibrium size and frequency of the polymorphic population. Although some paths lead rather directly to the equilibrium point, many others approach it in a more devious fashion. In general, polymorphic populations that are too large contract rapidly while converging on the equilibrium point; those starting with 40%–50% *A* tend to lose *A* momentarily before regaining it. Populations that are below equilibrium size approach the equilibrium point by relatively large gains in size or in frequency of *A* depending upon their starting gene frequencies.

The two charts of Figure 9 share a common axis, population size. Figures 10 and 11 represent the same two charts as one side and the "floor" of three-dimensional figures. In the case of the three-dimensional drawings, a new side—fitness-gene frequency—is available for depicting still further complications in the action of natural selection.

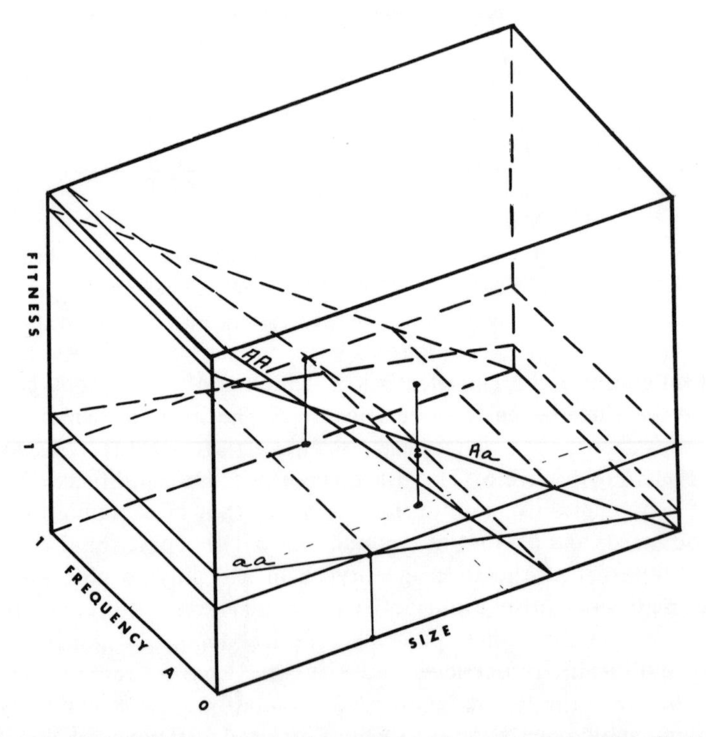

Figure 10. A three-dimensional drawing incorporating the relationships illustrated in Figure 9. Although various gene frequency-fitness relationships could have been illustrated in this drawing, the figure illustrates fitness that is independent of gene frequency.

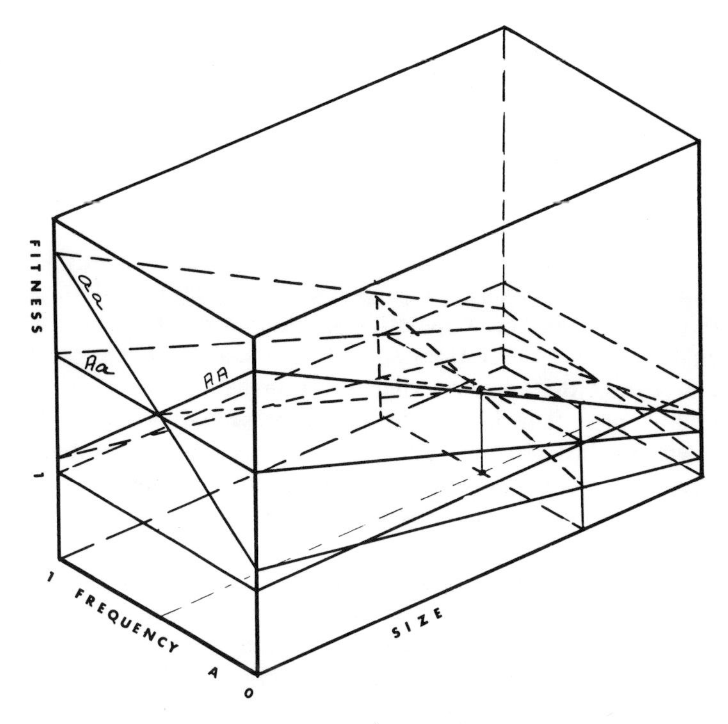

Figure 11. A three-dimensional drawing illustrating both gene frequency-fitness and population size-fitness dependence relationships.

The relationship between fitness and population size shown in Figure 9 was not affected by gene frequency. Precisely the same relationship as that shown in the two charts of Figure 9 has been reproduced in three dimensions in Figure 10. The absence of an effect of gene frequency on selection is shown by the lines of intersection formed by the back face of the solid and the tilted planes that represent the fitness of the different genotypes; these lines of intersection are parallel to the top and bottom of the back face.

The diagram illustrated in Figure 11 shows gene frequency-dependent, density-dependent fitness. The gene frequency-fitness relationships illustrated in this drawing are very simple ones:

Genotype	*AA*	*Aa*	*aa*
Fitness	1.5–p	1.0	1.5–q

The population reaches an equilibrium in respect to gene frequency when $p = q = 0.5$. Furthermore, it reaches an equilibrium in respect to population size when the intersection of the three planes representing the fitnesses of the three genotypes also intersects the plane representing an average fitness of 1.00. This and the preceding figure, complex as they may appear, represent very simple models indeed.

DISCUSSION

Birch (1960) has given a succinct account of the central problem of ecological genetics: ecology concerns itself with numbers of animals and what determines number. Population genetics concerns itself with kinds of animals and what determines kind. Ecological genetics concerns itself with the bearing the kind, or kinds, of animals have on their numbers, and of numbers on their kinds. The models presented in Figures 9, 10, and 11 deal with the last named relationships. They represent an attempt, however crude, to relate a number of observations in a meaningful manner and, in so doing, they deal exclusively with the interrelations of numbers and kinds.

The importance of density-dependent factors upon the Darwinian fitnesses of populations is clear: novel sources of premature death need not be met by a corresponding increase in the number of progeny produced. They are met, rather, by a reduction in population size. Population size, not progeny size, takes up the immediate shock of both genetic and nongenetic deaths.

Since alterations in population size bring about alterations in relative Darwinian fitnesses, the genetic load of a population cannot be used to predict the extinction of a population. The load concept is a static one based on equilibrium conditions; extinction is a dynamic process involving change in population size and, thus, in fitness relationships. Paradoxes arise if this fact is overlooked. Under the concept of genetic load as it has been defined, a calculated mean survival of 10^{-3} may appear too small to allow a polymorphic population to sustain itself; nevertheless, the same concept is unable to pass judgment on the ability or inability of a monomorphic population to exist when identical calculations show that its mean fitness would be many times smaller.

Throughout this article the word "fitness" has been used to mean Darwinian fitness, the relative ability of different individuals or different genotypes to leave fertile, adult offspring. It has been emphasized repeatedly that, for a population of constant size, the average number of daughters per mother—the mean Darwinian fitness of the population—must equal 1.00. There is still another "fitness"; that of the population. Has the model on the density-dependent aspects of Darwinian fitness helped in measuring population fitness in any meaningful way? Certainly not in an absolute sense! No definition of population fitness can be satisfactory in an absolute sense because the environment lies outside the scope of any such definition. Although it lies beyond the scope of the definition of fitness, the environment is very much involved in determining the fate of a population.

Despite the inadequacies that accompany it as a measure of population fitness, "size" as we have used the term is certainly involved. An amelioration of environmental conditions leads to an increased equilibrium population size (Figures 4 and 8). The continued existence of a population is more likely under ameliorated than under more severe environmental conditions. Thus, genetic changes which lead to an increased equilibrium size of a population can be regarded as changes that increase population fitness. [It is necessary to guard carefully against confusing an increase in equilibrium size with: (1) a fortuitous increase in numbers that merely reduces the productivity of surviving females; or (2) an increase in fecundity which may, in fact, result in a smaller equilibrium population size as shown by Nicholson and Bailey, 1935.] Genetic load, as we have seen, is unrelated to population size and, consequently, to the fitness of a population in the sense we are now discussing.

SUMMARY

An attempt has been made to incorporate density-dependent factors into genetic load theory. The adaptive values of individuals of different genotypes are measures of relative Darwinian fitnesses. However, a population at equilibrium size must have a mean Darwinian fitness (mean number of adult daughters per mother) of 1.00. Thus, adaptive values can be related to population size (Figures 4 and 5). The important point that emerges is that both genetic and non-genetic loads are absorbed not only by progeny size but by population size as well. Furthermore, with changing population size, the distinction between genetic and environmental loads is to a large extent removed; one type of load can be exchanged for another.

LITERATURE CITED

Beardmore, J. A., Th. Dobzhansky, and O. Pavlovsky, 1960, "An Attempt to Compare the Fitness of Polymorphic and Monomorphic Experimental Populations of *Drosophila pseudoobscura*," *Heredity*, 14:19–33.

Birch, L. C., 1955, "Selection in *Drosophila pseudoobscura* in Relation to Crowding," *Evolution*, 9:389–399.

———, 1960, "The Genetic Factor in Population Ecology," *Amer. Natur*, 94:5–24.

Crow, J. F., 1958, "Some Possibilities for Measuring Selection Intensities in Man," *Human Biol.*, 30:1–13.

Dobzhansky, Th., 1947, "Genetics of Natural Populations. XIV. A Response of Certain Gene Arrangements in the Third Chromosome of *Drosophila pseudoobscura* to Natural Selection," *Genetics*, 32:142–160.

______, 1948, "Genetics of Natural Populations. XVIII. Experiments on Chromosomes of *Drosophila pseudoobscura* from Different Geographic Regions," *Genetics*, 33:588–602.

______, 1950, "Genetics of Natural Populations. XIX. Origin of Heterosis Through Natural Selection in Populations of *Drosophila pseudoobscura*," *Genetics*, 35:288–302.

______, 1967, On Some Fundamental Concepts of Darwinian Biology. (manuscript)

Haldane, J. B. S., 1953, "Animal Populations and Their Regulation," *New Biol.*, 15:9–24.

Kimura, M., and J. F. Crow, 1964, "The Number of Alleles That can be Maintained in a Finite Population," *Genetics*, 49:725–738.

King, J. L., 1967, "Continuously Distributed Factors Affecting Fitness," *Genetics*, 55:483–492.

Lewontin, R. C., and J. L. Hubby, 1966, "A Molecular Approach to the Study of Genic Heterozygosity in Natural Populations. II. Amount of Variation and Degree of Heterozygosity in Natural Populations of *Drosophila pseudoobscura*," *Genetics*, 54:595–609.

Milkman, R. D., 1967, "Heterosis as a Major Cause of Heterozygosity in Nature," *Genetics*, 55:493–495.

Nicholson, A. J., 1955, "Density Governed Reaction, the Counterpart of Selection in Evolution," *C. S. H. Symp. Quant. Biol.*, 20:288–293.

Nicholson, A. J., and V. A. Bailey, 1935, "The balance of animal populations," Part I. *Proc. Zool. Soc. London*, Part III: 551–598.

Sved, J. A., T. E. Reed, and W. F. Bodmer, 1967, "The Number of Balanced Polymorphisms That can be Maintained in a Natural Population," *Genetics*, 55:469–481.

Wright, S., and Th. Dobzhansky, 1946, "Genetics of Natural Populations. XII. Experimental Reproduction of Some of the Changes Caused by Natural Selection in Certain Populations of *Drosophila pseudoobscura*," *Genetics*, 31:125–156.

-8-

Adaptedness and Fitness

THEODOSIUS DOBZHANSKY
Rockefeller University
New York, New York

In this article an attempt is to be made to disentangle, and hopefully to clarify, the concepts of adaptedness and of fitness, which are among the basic concepts of organismic biology. That living beings are adapted to their environments was recognized since John Ray, or even since Aristotle. A naturalistic explanation of the fact of adaptedness was suggested by Darwin. Darwin argued convincingly that adaptedness is a product of evolution controlled by natural selection, and he accepted Herbert Spencer's description of natural selection as "the survival of the fittest." Darwin's explanation of the origin of adaptedness is now pretty generally accepted as valid. It does not, however, follow that the Darwinian fittest, or at least the fit, are always and necessarily superior in adaptedness to the less fit. The relationships of adaptedness and fitness are often complex, making it advisable to keep these concepts separate.

IS ADAPTEDNESS A TRIVIAL MATTER?

The statement that organisms are adapted to the environments in which they live and reproduce may seem to be trivial. Adaptedness is the ability to live and to reproduce in a given range of environments. To some "modern" biologists this may seem to close the issue. A better acquaintance with living beings shows, however, that organic adaptedness is a far from trivial matter. There exist at least two million biological species, each with its own kind of adaptedness; many forms of life inhabit most extraordinary environments, or exploit their environments in most extraordinary ways. Perhaps the best way to convey my meaning is to consider some concrete examples.

The fly family Ephydridae contains about 1,000 described species. They live in such habitats as sewage, cadavers, sweet, brackish, and salt waters, including waters with very high salt concentrations. One species, *Psilopa petrolei*, has, however, managed to become adapted to live in pools of crude oil in the oil fields of southern California. As shown by Thorpe (1930), the larvae feed on other insects trapped in the oil, while these larvae themselves not only live submerged in the oil except for their spiracles, but their gut contains oil too. The adult fly can walk on oil surfaces as long as only the tarsi of its feet are in contact with the oil, but it becomes just as hopelessly trapped and killed as any other insect if other parts of its body are covered with oil. One of the fascinating aspects of the *Psilopa petrolei* case is that the habitat to which it is exclusively confined was evidently very uncommon, at least until the Standard Oil Company arrived in California. To be sure, saber-toothed tigers and other extinct animals occasionally came to grief in somewhat similar habitats; nevertheless, the evolutionary development of *Psilopa petrolei* must in all probability have been an extremely local operation. Anti-evolutionists may, I suspect, argue that such a development is simply inconceivable; but then they will have to suppose that God was in a very jocular mood when he decided to create this denizen of crude oil seepages in California.

A number of species in several genera of orchids have so modified the shapes and colors of some parts of their flowers that they resemble the females of certain species of bees or wasps inhabiting the same geographic areas. The resemblances are evidently close enough to mislead the males of the respective bee or wasp species, which attempt to copulate with the objects deceptively similar in appearance to their normal mates, and in so doing accomplish the pollination of the orchids. As Stebbins (1950) has pointed out, the resemblance is secured in entirely different ways by different species of orchids, and yet it is so contrived that each orchid species attracts the males of only a single insect species. The result is not only a pollinating mechanism but also a reproductive isolating mechanism which prevents gene exchange between species. The reliance of the plant on one sex of a single species of insect to accomplish the pollination may seem to be an extremely risky arrangement, but to quote Stebbins again "Orchids are very long-lived as individual plants, so that failure of seed formation in any one year is no great detriment to them. Furthermore, the number of seeds produced in a single capsule is perhaps the largest in the entire plant kingdom so that the number of successful pollinations required is less than in most other families." Yet, what biologist would have imagined this bizarre pollination mechanism before it was in fact discovered?

Somero and DeVries (1967) have published a brief account of a study of the temperature tolerance in three species of Antarctic fish of the genus

Trematomus. They occur in waters of extremely low but constant temperature, the average being $-1.9°$ C and the seasonal and depth fluctuations of the order only of $0.1°$ C. The tolerance range proves to be impressively narrow; the "upper incipient lethal temperature" is $6°$ C, which the authors believe is the lowest reported for any organism. Supercooling to $-2.5°$ is tolerated only if no ice crystals are formed, their formation causing rapid death. Warming up to $15°$ C brings death in six to eight minutes.

There are at least three reasons why the statement that anything alive is adapted to be alive does not, by itself, adequately describe the relevant phenomena. First, though all living organisms are evidently adapted to some environments, different organisms are adapted to different environments. Second, the adaptedness of individuals and the adaptedness of populations do not invariably go hand in hand. And third, although the problem of quantification of adaptedness has not been fully resolved, ways to achieve this goal seem to be in the offing. The adaptedness of some forms of life is clearly superior or more versatile than that of others.

ADAPTEDNESS OF INDIVIDUALS AND OF POPULATIONS

Adaptedness is a status of being adapted, i.e., of being able to live and to reproduce in a given environment. Adaptation is the process of becoming adapted. Adaptability is an ability to become adapted to a certain range of environmental contingencies. An adaptive trait, sometimes also referred to as an "adaptation," is a feature of structure, function, or behavior of the organism which is instrumental in securing the adaptedness (Dobzhansky, 1956).

It is evident that one cannot be adapted in the abstract, but only to a certain environment or a certain range of environments. Different organisms are adapted to different environments. My epidermis contains no chlorophyl or other photosynthetic pigment. Hence the old rhyme, "I cannot make a bun by simply sitting in the sun." Many plants capture the energy of sunlight and incorporate it into substances some of which go into making buns, which I eat to replenish my energy store. I am, therefore, adapted to pilfer some of the energy gathered by chlorophyl-containing plants. Zoology and botany have for at least two centuries pursued the research needed to understand the ways and means, as well as the limits of adaptedness of different organisms.

It is also evident that an organism must survive if it is to reproduce, and that it must reproduce if its offspring is to survive in the next generation. There exist adaptedness, adaptation, and adaptability on the individual level and on the population level. Individuals must be able to live, grow, develop, and mature. Mendelian populations and other aggregates of related organisms must reproduce to insure perpetuation of their kind. In general, individual and population level adaptednesses go hand in hand. Yet occasionally the two seem to be in conflict; situations when the welfare of individuals seems to be sacrificed for the benefit of their progeny are not rare.

For a Mendelian population, or even a clone, to be perpetuated it may be sufficient to have a single surviving pair or a single individual. The rest may be sacrificed. Thus, when a bacterial culture containing billions of cells is exposed to a lethal concentration of an antibiotic, such as streptomycin, the presence of a single resistant mutant cell is sufficient to perpetuate the strain. Populations of higher organisms may not be able to withstand so drastic a pruning; here the fate of an individual is hedged as much as possible against the dangers which occur with appreciable frequency in the habitats of the species. This is, however, a source of ambiguity in the phenomenology of adaptedness.

Individuals are protected, and yet they may be expendable. Among social insects the reproductive potential of sometimes a vast majority of individuals in a nest is sacrificed for the benefit of the colony, and the "workers" face dangers and death with what to a human observer looks like unbounded altruism and gallantry. Alas, this "altruistic" behavior lacks a possibility of choice, and hence is not altruistic at all. Perfect individual adaptedness would make individuals immortal. This is not fantastic, because some organisms have almost achieved just such a perfection. Some species of trees, of which the Sequoias are examples, not only live for centuries and millennia, but they are capable of stump-sprouting. An individual may, therefore, remain alive as long as physical and biotic environmental factors do not change drastically. Yet the Sequoias are relicts in danger of extinction; the high adaptedness and the near-immortality of the individual does not insure permanence of the species. The range of environments in which Sequoias flourish is a limited one; the high adaptedness is not matched by a high adaptability.

I can do here no more than merely mention the problem which logically presents itself at this point—that of senescence and death. Senescence and death are the antitheses of individual adaptedness, and attempts to find an explanation for their near-universality and evolutionary persistence have thus far met with only indifferent success. The idea that the old must be removed to clear the way for evolutionary innovation is too teleological, in the bad sense of the word "teleological," to be an acceptable explanation. More satisfactory is the consideration that genetic changes which increase the viability, vigor, and the reproductive potential of the young will confer a higher Darwinian fitness and will be selected even if they are detrimental to the old. Natural selection being opportunistic, the process of adaptation may result in improved adaptedness before and during the early stages of the reproduction, at the price of a disaster later on.

As a general explanation this does not seem to be entirely satisfactory. A genotype that would confer both a high viability and a high reproductive efficiency on its carriers early as well as late in life should have a Darwinian fitness superior to one that benefits only the young age. An ideal adaptedness would combine an indefinite prolongation of life with unlimited prolongation of undiminished reproductive ability. Very old Sequoias still bear numerous seeds. One has to suppose that for some reason genotypes advantageous at all ages arise less often than those which benefit the younger age groups alone. Evolution has contrived perhaps the best possible, but certainly not the best conceivable world.

TOWARDS QUANTIFICATION OF ADAPTEDNESS

Adaptedness is, in principle, measurable; in practice, the measurement is extremely difficult. Satisfactory methods of measuring the adaptedness, especially the adaptedness of populations, are yet to be devised. Awkward questions are sometimes asked, even by some biologists: In what sense can mankind be said to be better adapted than a fly or an alga or a bacterium? Are the so-called higher organisms in general any better adapted than the lower ones? If not, then was not the whole biological evolution a wasted effort? These questions are not easily answerable in precise terms, and our inability to give such answers attests the unsatisfactory state of our understanding of the phenomena of adaptedness and of their roles in evolution. The point is that the adaptedness of mankind is not unambiguously comparable to the adaptedness of a species of flies, or of algae, or of bacteria. These organisms exploit very different environments in different ways. The process of evolutionary adaptation is concerned not with adaptedness alone but also with adaptability. It can plausibly be argued that since flies can move from place to place, and thus seek and find suitable environments, while bacteria and algae can be transported only passively, the former have an adaptability superior to the latter. Man is certainly superior in adaptability, because he can create environments deliberately, according to plans devised by himself.

Some organisms can live in many environments, in different climates, and feed on many kinds of food; others are specialized for a much narrower range of environments. Some exploit very widespread environments, others (like the petroleum fly) an environment found only rarely and locally.

A measure of the adaptedness of a species or a population should, ideally, take all these facts into account. Such a measure might be obtained by summation of the particular adaptednesses over all the environments in which the population lives, times the frequencies of the respective environments. To the best of my knowledge, this kind of measurement has not been carried out.

A promising lead towards measuring adaptedness is the statistic named variously the Malthusian parameter, intrinsic rate of natural increase, or innate capacity for increase. The adjectives "intrinsic" or "innate" do not necessarily refer to the highest capacity for increase of which a species or a population is capable in some ideal environment. The ideal environment is not really known for any species, not even for man. The innate capacity for increase can be measured under a variety of conditions, favorable as well as unfavorable. What this parameter measures is the rate of growth of the population under a given set of environmental conditions, provided only that neither the food supply nor the space are limiting. This proviso is not unrealistic. Temporary relaxations of food and of space restrictions do occur in nature; in organisms which produce several generations per year, the populations may meet a superabundance of this sort in one or more generations, only to be cut down drastically during unfavorable seasons.

Birch and his colleagues in Australia, Lewontin, as well as our group at the Rockefeller University, have obtained some estimates of the innate capacity for increase in experimental populations of Drosophila. Let me mention a few of the results. Under relatively unfavorable conditions, chromosomally polymorphic populations of *Drosophila pseudoobscura* show higher capacities for increase than do chromosomally monomorphic ones. Ohba has shown, however, that under more nearly optimal conditions the difference diminishes or disappears entirely. Populations living for many generations in the artificial environments of the laboratory population cages undergo genetic adaptation to these environments. The improved adaptedness is reflected in higher values of the innate capacity for increase. It is, however, very instructive to observe that the improvement does not involve all the biological components but only some of them. Dobzhansky, Lewontin, and Pavlovsky (1964) found the improvements were brought about chiefly through increased fecundity of the young females, while their longevity not only failed to increase but actually declined.

This detail reflects, perhaps better than any theoretical discussion could, the limitations of the innate capacity for increase, determined in one or a few environments only, as a general measure of adaptedness. The longevity of the flies above a certain minimum is probably not very important in laboratory population cages. The numerical value of the innate capacity for increase is affected relatively little by variations in the longevity of the flies and by the fecundity of the older age classes; it is greatly dependent on the speed of the development, earlier onset of the reproductive period, and on the fecundity of young females. The relative importance of these variables under natural conditions may be quite different. The survival of the population in a local colony may easily be decided by the presence of a small number of exceptionally long-lived individuals, which survive unfavorable seasons and breed when favorable conditions return. The innate capacity for increase should, then, be determined by many environments as similar as possible to the natural ones. The labor required to do so would evidently be so great that such experiments have not been made.

DARWINIAN FITNESS

The basic postulate of the biological theory of evolution is that evolutionary changes are governed by natural selection. The environment presents challenges to which the species or a population responds by alterations of its gene pool. Natural selection acts as a cybernetic control mechanism, which transmits information about the challenges of the environment to the gene pool of the population. The classics of evolutionism described natural selection as the survival of the fittest. We prefer to describe it as differential perpetuation of genotypes or of genetic systems. Genotypes whose carriers differ in Darwinian fitness in a given environment are transmitted from generation to generation at different rates. This last statement is frankly tautological, and yet it is illuminating.

The Darwinian fitness (also called selective value or adaptive value) is not synonymous with adaptedness. As mentioned above, the adaptedness is, in principle at least, measurable in absolute units. The Darwinian fitness of a genotype is always relative to the Darwinian fitness of other genotypes. Darwinian fitness, like adaptedness, depends on the environment. The point is, however, that the Darwinian fitness is not necessarily predictable from information concerning adaptedness or vice versa. A genotype which is lethal in a given environment will, of course, have a zero Darwinian fitness. Yet a genotype with zero fitness may nevertheless be viable and fertile.

Experimental study of the chromosomal polymorphism in *Drosophila pseudoobscura* has yielded some instructive examples of imperfect correlation between the adaptedness and the fitness. In the populations of certain localities in California, third chromosomes with ST and AR gene arrangements are common and those with CH, PP, and TL gene arrangements are relatively rare. Rare gene arrangements occur in nature mostly in heterozygotes (heterokaryotypes) with other gene arrangements and seldom in homozygotes (homokaryotypes). Nevertheless, laboratory strains homozygous for any gene arrangement can be obtained and perpetuated indefinitely. Experimental populations monomorphic for any gene arrangement can also be maintained indefinitely in laboratory population cages.

Dobzhansky, Lewontin, and Pavlovsky (1964) showed that, in at least some laboratory environments, populations monomorphic for the gene arrangements which are common in nature usually show higher innate capacities for increase than do populations monomorphic for rare gene arrangements. Reference has already been made above to the fact that populations kept for a series of generations in laboratory population cages "improve" and show higher values of the innate capacity for increase. Interestingly enough, the most conspicuous improvements have been obtained precisely in populations monomorphic for rare gene arrangements, which to begin with showed the lowest capacities for increase. Biologically this makes sense: the chromosomes with rare gene arrangements are selected in nature mostly for coadaptation with other chromosomes present in the same populations, rather than for high fitness of the rare homokaryotypes. In our experimental populations the natural selection was given an opportunity to work for improvement of the homokaryotypes.

Experimental populations polymorphic for the ST, AR, CH, PP, and TL gene arrangements, taken two, three, four, or all five at a time, were kept in laboratory population cages. Here we could observe the changes in the relative frequencies of the chromosomes with different gene arrangements taking place from generation to generation and, using certain simplifying but not far-fetched assumptions, infer from the changes observed the Darwinian fitnesses of the different karyotypes (Pavlovsky and Dobzhansky, 1966, Anderson, Oshima, Watanabe, Dobzhansky, and Pavlovsky, in press). Some of the experiments were run in Dr. Oshima's laboratory in Japan, and others at the Rockefeller University in New York. In Japan, the ST/ST homokaryotype usually had the highest selective values, and the populations tended to establish high frequencies of ST chromosomes. In New York, the ST/AR heterokaryotype had the highest Darwinian fitness; and, as equilibrium states were approached, the AR chromosomes became equally or more frequent than ST. Most interesting of all, the homokaryotypes of the rare gene arrangements showed selective values close to zero; in fact, some estimates of the selective values turned out to be negative, which is a biological impossibility (negative values result evidently from sampling errors).

So, we are faced with a paradox: the carriers of the karyotypes which we know to have adaptedness sufficient to form reasonably flourishing populations by themselves have zero Darwinian fitness! A detail which is worth mentioning—larvae homozygous for the zero fitness karyotypes have actually been observed in experimental polymorphic populations when the gametic frequencies of the corresponding chromosomes were not too low. The paradox is removed if one realizes that a genotype (or a karyotype) has a Darwinian fitness not by itself but in relation to other genotypes. As mentioned above, the karyotypes which show higher adaptedness (as estimated from the innate capacity for increase) do tend to have also higher Darwinian fitness; there is, however, no simple proportionality between the adaptedness and the fitness.

Clatworthy and Harper (1962) have obtained illuminating data on the performance in pure and mixed laboratory cultures of four species of water plants belonging to the genera *Lemna* and *Salvinia*. The growth rates in the exponential phase (intrinsic growth rates) in pure cultures are in the order:

minor > *natans* > *gibba* > *polyrrhiza*.

The growth rates in self-crowding cultures (arithmetic growth rates) gave, however, the order:

natans > *polyrrhiza* > *gibba* > *minor*

The total dry weight achieved in pure cultures was in the order:

polyrrhiza > *minor* > *natans* > *gibba*

Pairs of species were now grown in mixed cultures. When *polyrrhiza* is grown with *natans* or with *gibba*, the proportion by weight of *polyrrhiza* declines continuously and tends eventually to disappear. In mixtures of *polyrrhiza* and *minor*, both species maintain themselves, although the authors found indications that *polyrrhiza* may ultimately be the victor. What then determines which is the successful species when in competition? The authors infer that the outcome of the competition is determined by certain morphological characters of the competitors, which enable the victorious species to occupy the surface layers of the mats of fronds formed in the experimental cultures, and thus avoid being shaded from light to which the cultures are exposed.

A biological situation in which the distinction between the adaptedness and the Darwinian fitness is disclosed most clearly is expansion and contraction of the population size versus stability or flexibility of the genetic composition of the population. In animals, such as Drosophila which produce several generations per year, the populations may be decimated during the cold or during the dry seasons, and expand enormously during the more favorable warm or wet seasons. Even in tropical countries, in which the temperature and moisture are sufficient the year around, some species of Drosophila become abundant or are reduced to scattered survivors when certain fruits in which they breed are available or absent. Now, a population which is shrinking in size does not necessarily have zero adaptedness under the environmental conditions which make it shrink. If winter became permanent or very prolonged, the Drosophila species of cool-temperate and cold countries would become extinct, but prolonged dryness in the tropics might still allow a reduced population to hold on in some less dry localities. It is, however, reasonable to regard a shrinking population as having a lower aggregate adaptedness to the conditions which make it shrink, and an expanding population as having a higher adaptedness to the environment which makes it expand.

We now inquire whether the Darwinian fitness oscillates hand in hand with the adaptedness. Some insight in this may be gained by examining a polymorphic population of a species, or a group of related and ecologically not too dissimilar sympatric species. In *Drosophila pseudoobscura*, cyclic seasonal changes in the relative abundance of different karyotypes have been detected in several localities in California; in other localities the changes are absent or too small to be easily detected. The European *Drosophila subobscura* shows relatively little evidence of seasonal genetic changes in its chromosomal composition. Where changes do occur, the Darwinian fitness of certain karyotypes is evidently increased, and of other karyotypes decreased, during the periods when the adaptedness of the population is high and when it is low. Where no changes occur, the Darwinian fitnesses remain unchanged in the face of the changes in the adaptedness. *Drosophila pseudoobscura* and *Drosophila persimilis* are sympatric and apparently rather similar ecologically in western North America. Their relative abundance in some localities where they were studied varies quite appreciably from year to year, but rather less from season to season. The causes of the variations in relative abundance are obscure; they are not reflected in the chromosomal composition of either species.

The environments which determine the adaptedness and the Darwinian fitness certainly include both physical and biotic components. An objection may then be made that the Darwinian fitness is merely a consequence of mutual modification of the adaptednesses of the genotypes or of populations. If so, some colleagues have contended, distinguishing adaptedness and Darwinian fitness is superfluous. This objection, I think, is not valid. True enough, examples are known when the Darwinian fitnesses, i.e., the relative selective values, of two karyotypes of *Drosophila pseudoobscura* are dependent on their relative frequencies, or on the presence or absence of still other karyotypes (Levene, Pavlovsky, and Dobzhansky, 1958). The mutant ebony and the wild-type of *Drosophila melanogaster* influence each other when living in the same culture (Weisbrot, 1966). However, the very fact that, as mentioned above, the order of adaptednesses of different karyotypes of *Drosophila pseudoobscura* in monomorphic populations generally parallels their Darwinian fitnesses when they are present in polymorphic populations militates against the supposition that the variations in adaptedness are simply products of mutual facilitation or interference of the karyotypes. One may, of course, speak of absolute adaptedness and relative adaptedness (= Darwinian fitness); but this does not seem to be conducive to much clarity of thinking.

ADAPTEDNESS, FITNESS, AND PERSISTENCE

In articles containing a most perceptive analysis of the "Components of fitness," Thoday (1953, 1958) defined fitness as the "probability of leaving descendants after a given long period of time. Biological progress is increase in such fitness." Fitness in Thoday's sense is evidently not identical with either adaptedness or Darwinian fitness. It may perhaps be referred to as persistence. To be operationally useful, adaptedness and Darwinian fitness must explicitly refer to the time and to the environment, or the range of environments, when and where they are measured. The evolutionary persistence is operationally elusive; it depends on the countless contingencies which a species or a population will be facing in the future, and Thoday specifies a remote future, such as 10^8 years! At least given the present level of knowledge and techniques, there seems little chance of predicting the course of evolution even over time intervals some orders of magnitude smaller than indicated by Thoday.

The persistence, or the "fitness" in Thoday's sense, is nevertheless something no theory of evolution can afford to ignore. I agree with Slobodkin (in the present volume) that the best evolutionary strategy is that which maximizes the chances of persistence. The prime dichotomy in evolution is survival vs. extinction, life vs. death. All organisms living today are presumed to be descendants of the primordial life, the date of the origin of which has been pushed back by recent discoveries to three billion years ago or even longer. However, innumerable collateral relatives of the organisms now extant have become extinct. Those which survived did so in two ways. First, they underwent countless genetic changes which made them radically unlike their ancestors. Second, some survivors clung tenaciously to their old ways of life and presumably to their genetic constitutions. These are, of course, not "either-or" alternatives, since different degrees of changeability and stability are encountered.

The so-called "living fossils" are forms of life which have neither changed nor become extinct for long periods of geological time. Barghoorn and Tyler (1965) described a microfossil *Kakabekia umbellata* from a geological deposit estimated to be two billion years old. Siegel *et al.* (1967) found what looks like the same organism alive in the soil heavily polluted with human urine near the walls of a castle in Wales. This organism can be grown in laboratory cultures with 50 per cent ammonia in the air. *Kakabekia umbellata* evidently has an adaptedness good enough to survive in an extremely specialized habitat. What is its Darwinian fitness? The question is almost meaningless. One may, of course, imagine that the *Kakabekia* used to live in other habitats, and was displaced by competitors possessing a superior Darwinian fitness; one may also imagine that in the foul soil near its Welsh castle *Kakabekia* is naturally selected in preference to other competitors. Anyway, *Kakabekia* does possess an adaptedness to live in an extraordinary habitat, at least in the laboratory. One may speculate that it is a survivor from most remote ages, when the earth's atmosphere contained ammonia and other gases which gave rise to the organic compounds from which life arose. If so, the ecological niche of *Kakabekia* may have been far more widespread than it is now.

The line of descent leading to man is contrasting spectacularly with that of *Kakabekia*. There was no animal resembling man until two million years ago at most. The remains of the race *habilis* of *Australopithecus africanus* (otherwise styled *Homo habilis*) are perhaps 1.7 million years old. The evidence appears to be conclusive that this animal made primitive stone tools. Farther back, in mid-Tertiary, there were forms which were perhaps the common ancestors of the anthropoid apes and of man. In early Tertiary there were primitive primates not yet radically different from the insectivores. The origin of mammals took place in the Mesozoic from reptile-like ancestors.

The ancestors of *Homo* had what Thoday designates as "genetic flexibility"; the ancestors of *Kakabekia* either had little of such flexibility, or else had, in at least the line of descent leading to the new living form, no opportunity to make use of it. Both have survived. This surely does not mean that *Homo* and *Kakabekia* are equally successful products of the evolutionary process. *Kakabekia* possesses an adaptedness to live in a very narrow ecological niche; *Homo* is able to widen the range of his living opportunities by manipulation of his environments. So there may be forms of the evolutionary process which deserve being called "progressive." This topic is, however, outside the frame of reference of the present article.

LITERATURE CITED

Anderson, W. W., C. Oshima, T. Watanabe, Th. Dobzhansky, and O. Pavlovsky, 1968, "Genetics of Natural Populations. XXXIX. A test of the possible influence of two insecticides on the chromosomal polymorphism in *Drosophila psueodobscura*," *Genetics*, 58:423–434.

Barghoorn, E. S., and S. A. Tyler, 1965, "Microorganism from the Gunflint Chert," 2nd ed. *Science*, 147:563–577.

Clatworthy, J. N., and J. L. Harper, 1962, "The Comparative Biology of Closely Related Species Living in the Same Area," *J. Exp. Bot.*, 13:307–324.

Dobzhansky, T., 1956, "What is an Adaptive Trait?" *Amer. Natur.*, 90:337–347.

Dobzhansky, Th., R. C. Lewontin, and O. Pavlovsky, 1964, "The Capacity for Increase in Chromosomally Polymorphic and Monomorphic Populations of *Drosophila pseudoobscura*," *Heredity*, 19:597–614.

Levene, H., O. Pavlovsky, and Th. Dobzhansky, 1958, "Dependence of the adaptive values of certain genotypes in *Drosophila pseudoobscura* on the composition of the gene pool," *Evolution*, 12:18–23.

Pavlovsky, O., and Th. Dobzhansky, 1966, "Genetics of Natural Populations. XXXVII. Coadapted System of Chromosomal Variants in a Population of *Drosophila pseudoobscura*," *Genetics*, 53:843–854.

Siegel, S. M., and C. Guimarro, 1967, "On the Culture of a Microorganism Similar to the Precambrian Microfossil *Kakabekia umbellata* Barghoorn in NH -rich Atmospheres." *Proc. Nat. Acad. Sci.* 55:349–353.

Somero, G., and A. L. DeVries, 1967, "Temperature Tolerance of Some Antarctic fishes," *Science*, 156:257–258.

Stebbins, G. L., 1950, *Variation and Evolution in Plants*, Columbia University Press, New York.

Thoday, J.M., 1953, "Components of Fitness," Symp. Soc. Exper. Biol. 7:96–113.

———, 1958, "Natural Selection and Biological Progress," In *A Century of Darwin*, S. A. Barnett, (ed.), London, Heineman.

Thorpe, W. H., 1963, *Learning and Instinct in Animals*, 2nd ed., Methuen, London.

Weisbrot, D. R., 1966, "Genotypic Interactions Among Competing Strains and Species of Drosophila," *Genetics*, 53:427–435.

-9-

The Population Flush and Its Genetic Consequences

HAMPTON L. CARSON
Washington University
St. Louis, Missouri

INTRODUCTION

Populations of organisms are never truly stationary in size. Some numerical changes in population are essentially symmetrical; thus, the population rises from some level of stable equilibrium and then frequently falls back again to approximately its original size. Such changes have been referred to as oscillations (Allee *et al.* 1950). They are distinguished from what these authors call fluctuations by their regularity and symmetry. Both oscillations and fluctuations are distinguished from "spurts" or "crashes." The latter are defined as special cases in which the population departs radically from the normal pattern of equilibrium so that an excessive maximal fluctuation peak is attained.

The present discussion embraces all such shifts, although the emphasis will be placed on the radical departures. Profound changes such as the latter would be the type of change most conducive to the action of the genetic phenomena to be described. The ascending phase of population growth will be referred to as a "flush" and the subsequent fall as a "crash." These terms will serve to emphasize the point, which will be made later, that crucial alterations in gene pools are most likely to occur when the changes have the precipitate nature suggested by these rather dramatic terms.

Cyclical oscillations characteristically occur in populations of various organisms like lemmings, voles, grouse, or foxes. Because of the conspicuous nature of these animals or because of their economic importance, these phenomena have attracted an immense amount of attention. The subject of their causation has been extensively reviewed and discussed (e.g., Elton, 1942, 1958; Butler, 1953; Hewitt, 1954; MacArthur, 1955; Errington, 1957). According to Lack (1954) the basic cause of these cycles is probably not variation in a single climatic factor. Rather they appear to be the outcome of a predator-prey oscillation set off by periodic shifts in the vegetable food available, for instance, to a dominant rodent. Even in the clearest of these cyclic changes, however, the contributing factors are multiple, and their causes are buried deep within the complexities of community ecology.

With respect to the present discussion, however, it does not matter what the causal agents of an increase in population size are. What is most important in the present context is the simple fact that a population flush has occurred, taking place within relatively few generations. The present paper concerns itself principally with a genetical topic; namely, the reorganizational changes in the gene pool of an organism which accompany the rise and subsequent fall of population size. A penetrating review of pertinent literature dealing with this subject will be found in Ford (1964).

BIOLOGICAL PRECONDITIONS TO FLUSHES OF EVOLUTIONARY IMPORTANCE

To be genetically important, population size shifts do not have to be extermely large. A 10 or 20 per cent increase might serve to set off some of the changes to be described below, but unfortunately, data on this point are lacking. To consider an example, even the least variable of the four species of European tits studied by Kluijver (1951) showed at least one doubling of the breeding population size over a twelve-year period. Less well-documented cases could be multiplied.

Most population flushes are temporary and a decline or crash usually follows them. The considerations in the present context, furthermore, are expressly confined to populations that maintain a gene pool wherein one generation of sexual reproduction follows rapidly upon another. The group concerned must be sexually reproducing and cross-fertilizing. The concepts developed here would thus not apply to clones or any other aggregations or groups of apomicts or asexually-produced entities. Such conditions lack the properties of a gene pool and would not necessarily follow the pattern about to be described.

Not only are vegetatively reproducing or clonal organisms to be excluded, but for simplicity, the principles advanced here will be considered only as they apply to those organisms which are diploid and have extensive genetic variability which is free to recombine during the flush. Organisms which show population excrescences based primarily on facultative apomixis, with only an occasional sexual generation, are also excluded. Although some of the principles advanced here might also apply to them, such downgrading of the sexual mode of reproduction has a dampening effect on all evolutionary processes, including those to be considered here.

The writer (1957) has referred to those organisms having relatively high chromosome numbers, relatively high chiasma frequencies in both sexes, as well as a well-developed capacity for outcrossing, as having "open" recombination systems. Polyploid organisms, moreover, do not have very open systems; they suffer considerable reduction of their capacity to release genetic variability relative to diploids. This statement, of course, should not be taken as including those organisms which have polyploidy in their remote ancestry and which have become secondarily diploidized by a more recent evolutionary development.

The genetic system that we are considering here, furthermore, must not have its outcrossing mechanisms impaired by obligatory or even frequent self-fertilization or similar impairment of the outcrossing system. On the other hand, no actual bars to inbreeding through sib matings should be present. A facultative inbreeder-outbreeder is ideal. Outbreeding prior to a flush would be especially conducive to the development of the conditions deemed important by the writer.

I also wish to exclude those organisms which have adopted, apparently through compromises in their past evolutionary history, inordinately long periods between sexual generations. Thus, although *Sequoiadendron giganteum* may qualify in other ways to be included in the select company here described, it has sacrificed, relative to many other organisms, the capacity to move fast in evolution because of the great time between generations.

Almost all the above requirements would be expected to contribute to the existence and maintenance of freely recombining genetic variability in the population of the species in question. This variability may, of course, be concealed in heterozygotes. On the other hand, variability held within non-recombining chromosomal supergenes, such as inversions or translocations existing in balanced heterozygous form, is less important. Availability of such variation to release through recombination is limited.

The organism whose flushing populations we are going to trace, therefore, has the following characteristics. It reproduces almost exclusively by sexual means, is dioecious, and normally outcrossed. In animals, the members of both sexes are not sedentary but move freely in the population, assuring random mating within a large group of organisms. In plants the burden of this function is, of course, borne by wide pollen dispersal. The organism is diploid and has a relatively high chromosome number. Chiasma frequency is high in both sexes. The generation time is short and such that should an alteration in environment provide an ecological opening, the opportunity to expand can be followed by a rapid numerical increase. This would involve a number of sexual generations and not merely an apomictic or vegetative response. The innate capacity for increase in numbers of the organism (Andrewartha and Birch, 1954) should be substantial. The dispersal mechanisms of the organisms must be adequate to carry a number of the individuals produced during the flush outside of the home territory. The organism should have enough adaptability so that the somewhat foreign environment which migrant individuals may reach is not immediately lethal to them. Finally, the organism should not begin its flush with a depauperate gene pool.

The formulation to follow is predicated on the assumption that the composition of the gene pool, and the genetic system which manipulates it, have a profound influence on the evolutionary future of the descendants of a population which undergoes a flush and crash. Essentially what will be described is a type of breakdown in the genetic basis of fitness which accompanies the period of reduced natural selection during the period of flush. The crash proceeds precipitately because the existence of stringent environmental conditions and genetic breakdown products coincide in time. As the crash reduces the population size towards the original level, it may carry the population to a size even smaller than at the inception of the cycle. This may be because of the lag between the appearance of stringent conditions and the multiplication by selection of individuals carrying genotypes with greater Darwinian fitness relative to these new conditions.

GENETIC CHANGES PRIOR TO A POPULATION FLUSH

Although concern with the ultimate cause of population increases and decreases was disavowed in earlier paragraphs, it may be well to emphasize that under some circumstances a purely genetic change may be the stimulating event for a population flush. This has been demonstrated in experimental populations of *Drosophila melanogaster* (Carson, 1961; Cannon, 1963). An inbred population marked by recessive mutants has a size (161.3 ± 7.3 individuals) which is roughly equilibrated with its food source, the latter being held constant over many generations. The population is altered genetically by adding a single male fly whose mother was from the population and whose father is from an unrelated wild-type laboratory strain. Following the infusion of this new genetic material, a population flush ensues. This occurs without change in food resources or space; the population size increases by a factor of about three (477.3 ± 11.7 individuals) in approximately nine generations. Even though the food is unchanged in these experiments, population growth is exponential. This rather remarkable change is evidently due to the higher

Darwinian fitness of at least some of the genotypes produced following the initial hybridization. Thus, these individuals share in a new gene pool which, relative to the original one, sustains a larger size.

In these experiments, moreover, the population reaches a peak size and then falls abruptly, oscillating around a level somewhat below the original peak. Still later, after the passage of 30 to 40 generations, the population sinks lower, until its size is only three-quarters as large as its earlier level. These events, as observed in these experiments, appear to parallel the rise and fall of populations in nature, although in the genetically-induced flush in laboratory populations, the rise appears to have been a permanent one and no real crash follows it. In discussing genetical change as a causal factor, it is not intended to argue that such hybridizations are a common cause of population flushes under natural conditions. Probably most such increases are caused by the more familiar alteration of some key ecological condition, such as a new availability of excess food. Under laboratory conditions, of course, such flushes may be demonstrated with great ease. Because of this, most ecologists tend to discount the hybridization possibility and, under natural conditions, proof of hybridization prior to a flush is very hard to come by. There are, nevertheless, a number of cases which are amenable to such an explanation, such as that of the fire ant (*Solenopsis saevissima*) in the southern United States (Wilson and Brown, 1958) and many instances of natural hybrid swarms (see Stebbins, 1950).

This discussion was begun with the mention of hybridization as providing a genetic basis for a population flush for a particular reason. If an ecologically-caused flush begins with a depauperate or relatively invariable gene pool, there will be little difference between the genotypes during flush, crest, crash, and post-crash periods. Indeed, if the population returns to the same size as the original, the composition of the gene pool may have remained unchanged throughout the entire cycle. Accordingly, no real evolutionary event will have occurred.

This consideration emphasizes a point made earlier, namely that a population flush and crash cycle can only have evolutionary significance if the gene pool or the subdivisions formed from it contain substantial genetic diversity.

Many small populations, poised on the threshhold of a flush, may well have such genetic variability wholly stored within them. On the other hand, if a hybridization occurs and is followed immediately by a flush, an even greater release of variability may be expected. Thus, should a period of hybridization immediately precede an ecological opening in the form of increased food supply, for example, the magnitude and momentum of the flush finds itself augmented by both genetic and ecological factors simultaneously. It is possible that some especially great population flushes draw on these two elements simultaneously. Clearly, the human population of the world, since the fourth millenium B.C., has travelled this kind of precipitous course, and both outcrossing by hybridization and ecological permissiveness have played a role. In this case, as in most natural ones, it is not possible to assess the role of the outcrossing process apart from ecological opportunity.

While considering those genetic changes which might occur before the actual flush begins, it should be pointed out that, generally, seasonal cycles or breeding season flushes are of too short a duration to function in an efficient manner from the evolutionary point of view. Although the rise and fall of a population with season appears to fulfill the conditions being set down here, there are several reasons why the writer feels constrained to look beyond the seasonal to departures of greater magnitude. In the first place, many organisms displaying seasonal change seem to be caught up in a series of obligatory balanced conditions, with the seasonal genetic change not being a drastic or far-reaching one genetically. Frequently, the gene pool may be observed to return to precisely the same condition as was observed at the same time during the previous season.

Examples of such populations which fluctuate in size and display concomitant dynamic genetic equilibrium are the well-known seasonal cycles in *D. pseudoobscura* (Dobzhansky, 1948; Epling, Mitchell, and Mattoni, 1957), in *D. robusta* (Levitan, 1951), in *D. funebris* (Dubinin and Tiniakov, 1945) and in *D. melanogaster* (Band, 1964; Oshima and Kitagawa, 1963). In all of these cases, study of chromosomal inversion polymorphism or the frequency of lethal genes from year to year show that the polymorphism observed initially is either maintained in essence or replaced by an equivalent polymorphic system. This is not to say that a population is exactly the same as it was in a previous year; indeed in some instances small differences are apparent (see Dobzhansky, 1963). Nevertheless, the organisms mentioned above are highly dependent on balanced polymorphism. In these cases recombination is frequently inhibited by low chromosome numbers, inversions, or both. Dependence on balance and heterozygote superiority in these species frequently means that the recombinant types formed during the seasonal flushes are wiped out during the seasonal crashes. The population is thus forced back again to dependence on the same (or perhaps slightly improved) balanced heterotic systems.

Quite apart from the genetic system of the particular organism concerned, however, the seasonal cycle is not considered to be very conducive to the major genetic reorganizations that are being stressed here. Most seasonal cycles proceed in an orderly manner in a highly adapted and integrated ecological community. Thus, diseases, parasites, predators, and reduction of food press in upon and dampen the expanding population even as it begins. In other words, selection pressures are continuous throughout. A much more important type of population flush is one which extends over three or four seasons in the temperate zone, for example. Such an event must indeed be deemed relatively rare when compared with the simple seasonal flush. The absence of the repeated obligatory oscillations associated with seasons in the tropics sets the stage for the occasional grand-scale flush that is deemed to be especially important. This statement should not be construed as a contention that seasonal cycles do not occur in the frost-free tropics. Many tropical organisms do show strong seasonal cycles. Nevertheless, the dampening hand of winter on the evolutionary process is a fact of life in the temperate zone which affects all populations within its reach. Freedom of the tropical biota from such a stringent seasonal restriction places many species in the position wherein they may avail themselves immediately of both genetic change through hybridization or ecological change, whichever should happen along at any season of the year.

GENETIC BREAKDOWN DURING THE LOG PHASE OF THE FLUSH

A population flush must begin by the survival of a greater number of zygotes to the reproductive stage than previously. In genetic terminology it may be said that the inception is represented by the rearing of a large F_1 under conditions wherein natural selection for survival is less than normally encountered. This relaxation of selection is continued and the next step is the production of an F_2, F_3, F_4, and later generations still under conditions of relaxed selection. The conditions are favorable so that the adult females do not suffer, as they normally do, from food shortage feedback which affects their fecundity. The innate capacity for increase is dramatically realized and released, perhaps for the first time in many thousands of generations.

One of the most striking features of this period, genetically speaking, is what may be referred to as a breakdown in the genetic basis of fitness. The work of Vetukhiv (1954) and Brncic (1954) have documented what happens when hybridizations are made between populations. There is in the F_1 a manifestation of heterosis, but in the F_2 and later generations this heterosis disappears as its genetic basis is broken down by recombination. Brncic, for example, concludes that an originally internally balanced gene complex may suffer disintegration owing to recombination and crossing over in the progeny of hybrids.

Although the phenomenon is perhaps most easily observed under conditions of hybridization (see especially Stebbins, 1959), the same process could occur in a population which has not been hybridized but nonetheless carries some genetic diversity. The removal of selection permits the release of this recombination which results in the production of actual living and surviving individual recombinants which swell the population beyond all previous levels.

THE CREST OF THE FLUSH

For the reasons which have been adduced earlier, the crest of the population flush is characterized by a multitude of recombinant genotypes. Under the conditions that prevailed during the log phase of growth, relatively few recombinants would have been subjected to selective elimination. Carriers of genes with drastic effects such as lethals and sterility genes, of course, would constitute exceptions which would not survive to the crest. Populations at the crest would accordingly be characterized by extraordinarily great polygenic variability including many recombinants with relatively low fitness. A striking case has been described by Ford and Ford (1930). Much of this variability would be expected to be cryptic, as most variability in natural populations is, but segregation of some recessive visible mutations might occur.

Density-dependent factors would be expected to come into play as the crest is reached. Food and space shortage and the dampening hand of selection would develop suddenly. High mortality then replaces the previous very low mortality, as competition for food and breeding space ensues.

One of the inevitable results is dispersal or migration from the affected area, a process that has been repeatedly described as characteristic of such population phases. There certainly need be no specific genetic predisposition to migration, but should such a state arise or be inherent among the recombinants, it could reinforce the ecological pressures leading to outmigration.

Inherent in such a situation, however, is an essentially random movement of genotypes. Those remaining behind to continue the population at the site of the flush would likewise be expected to be randomly chosen as to genotype, unless, indeed, they represent certain weak recombinants which are unable to migrate effectively.

Frequently, the crash is so precipitous as to carry the population from inordinate heights to an extremely low ebb (see e.g., Clough, 1965). A number of ecologists have sought explanations for such apparent overly precipitous drops (e.g., Chitty, 1957). It may be pointed out that genetic changes related to recombination and breakdown of fitness may indeed be involved in such instances. As pointed out elsewhere in this article, a rather long lag period of many generations would be experienced after genetic breakdown and before the population could reap the benefits of renewed natural selection (e.g., Haldane 1957).

In this connection, it may be noted that what has here been called the genetic breakdown might indeed operate independently from density-dependent factors. Although conditions of crowding might make the fitness of dysgenic recombinants lower, the phenomenon might well be strongly manifested in the absence of crowding and due, for example, to the effects of certain gene combinations on sterility. Accordingly, at the crest of the flush, the population is in a strangely vulnerable position genetically and such factors may well play a key role in the future adjustment and adaptation of the species.

The population of the species which remains resident in the geographical area where the main flush occurred will probably be sharply different in its recombinational makeup following the crest, even if no crash ensues. This would result from the chance effect among the key survivors. The changes would be analagous to those occurring following random drift. This point will be considered further in the subsequent treatment of the crash.

THE PHASE OF DISPERSAL

Movement out of the immediate area of the flush will carry the recombinant products of genetic breakdown into new territory. The migration waves may be concentric or they may follow somewhat skewed paths. Characteristically, however, under conditions of great population size, one may expect a large random ingredient in these movements. Many individual migrants move into areas which are wholly unsuitable to their survival and quickly die out. It is unlikely that sensory perception of proper habitats plays a large role among such migrants; they flood the adjacent geographical areas without regard to its suitability for their colonization.

Whatever the forces behind the process of dispersal, the very fact that it occurs has great importance for the evolutionary future of the group. This is primarily because the phase of out-migration has inherent in it a possibility of successful colonization of some new area by migrant individuals. These colonizations would furthermore be likely to be characterized by the "founder effect" (Mayr, 1954). As the out-migration proceeds, following the crest of the flush, one can conceive of the formation of a mosaic of new population foci. Many of these colonizations would be of a chance nature and might, for the first time in many thousands of generations, establish genetic contact between the new migrants and isolated enclaves or foci formed previously.

Such interbreeding would be expected to produce some small subsidiary population flushes of the kind which are mediated by heterosis. Although this provides further opportunity for genetic recombination, it is probably not very significant when compared to that resulting from the major flush.

Certain dispersants may reach areas previously not occupied by the species. Population foci started in such areas would be of utmost significance from the evolutionary point of view. These could provide classic examples of the power of genetic drift involving random changes in gene frequencies. The present discussion, moreover, emphasizes the highly recombinant nature of the genotypes of the founding individuals. Such recombinants, arising as they do from a great population flush, might indeed represent bizarre combinations of genes. Some loci might indeed show fixation relative to the parent population.

To further emphasize this point, it may be said that ecological dispersion and colonization leads directly to dispersion of another kind, namely that of gene frequencies. These effects have been repeatedly emphasized through a long series of papers by Wright (1964). The genetic dispersion which occurs has been brilliantly reviewed by Falconer (1960), and demonstrated in laboratory experiments by Buri (1956). Indeed, if the gene pool of the population remaining *in situ* may be subject to dispersive effects, then such events are even more characteristic of populations which are founded by migrants moving out from the area where the flush occurred.

THE POPULATION CRASH AND THE FORMATION OF DEMES

The subsiding of the flush and the re-institution of strong selection should indeed result in the elimination of a great multitude of genotypes. This process would affect the non-breeding population first and would have the result of producing a population situation wherein a number of new isolates would be formed. This phase would therefore produce a mosaic pattern of distribution although the demes produced, unlike the earlier phase, would not be in contact with one another, primarily because of the nonsuitability of the intervening area for the breeding of the species.

As the population decline intensifies, many of the breeding foci may turn out to be abortive colonizations; the individuals comprising them die out. There follows an opportunity for a new and powerful evolutionary force, namely, interdeme selection (see Lewontin, 1965).

It should be emphasized that, during the whole population cycle that has been described, there has been little chance for adaptive evolution. Selection is considered to have been primarily an eliminative process with little constructive ingredient with respect to adaptation.

In a manner analagous to intra-population elimination of weak or sterile recombinant individuals, many small local populations having dysgenic gene pools are also eliminated. According to this view, the surviving demes are the result of negative selection. By this is meant that a population survives not because of any superior adaptive property possessed by its individuals, but rather because its gene pool is free of highly dysgenic and severely deleterious properties. It should perhaps be further emphasized, as has been done by Williams (1966), that there are strong arguments that interdeme selection seems unlikely to be able to produce adaptation. If selection eliminates the bad demes, as indicated above, those remaining may survive not because the individuals comprising them have specific adaptations to the precise demands of the environment, but rather because survival may be due to individual adaptability through a general, physiological, or developmental homeostasis.

SPECIES FORMATION FOLLOWING THE POPULATION FLUSH

Certain demes which survive in isolation following population size recession would be likely to have new and unique gene pools for the reasons set forth earlier. Some of these genetic changes would be likely, again by chance, to modify reproductive properties in some way. Gene changes might affect structures or biochemical properties directly concerned with sexual reproduction, or they might directly affect the mating behavior in the case of animals. One of the first things that would happen in a deme would be for selection to mutually adjust the two sexes. This would probably be an active selective process even in a small population and might be responsible for the fixation of bizarre secondary sexual characteristics.

The features fixed in the genotype during this phase could serve as very strong initiators and perhaps cementers of reproductive isolation, so that when and if another population flush occurs, not all of the former isolates would be swamped out of existence. Such key reproductive isolations could result in the origin of new species which may quickly become sympatric with the older or "parent" species.

The extraordinary proliferation of species in such places as Lake Baikal, or on oceanic islands like Hawaii, may indeed be related to population flushes and chance effects following the subsequent crashes. The argument is furthermore advanced that much of the genetic change which results in the permanent cessation of gene flow is not related to the repatterning of the gene pool by selection for specific adaptations.

It follows from this that when first formed, allopatric synchronic species will differ primarily in genetic characteristics which are the outcome of the haphazard effects of population sampling on reproductive behavior as well as other chance fixations. Once this phase is past, however, the classical process of adaptive evolution by mutation, recombination, and selection resumes. Different adaptations may be forged by these processes and become fixed. As Haldane (1957) has pointed out, this process, if it is to involve many loci, will take a long time because of the genetic load imposed while each allele is in process of being fixed. No such load will prevail under the circumstances of the quick homozygosity which may be conferred on certain demes following a population flush and crash.

As a result of these processes, two closely related species are likely to differ in both non-adaptive and adaptive ways. The idea also accounts for the widespread phenomenon of sympatric species. Such species, in many instances, have similar basic adaptations and occupy extraordinarily similar habitats. For example, despite the very extensive study of the sibling pairs *Drosophila melanogaster-D. simulans* and *Drosophila pseudoobscura-D. persimilis*, the differences between the species, with regard to their major adaptations, appear trivial indeed. This statement, however, does not ignore the obvious fact that each fits into a slightly different niche; the exclusion principle of Gause (1934) certainly applies to them as to all other species. Two gene pools, such as those possessed by these very similar species, can never be identical and thus must of necessity exploit the environment in slightly different ways. These differences, however, may have had a fortuitous origin followed by a moderate strengthening after gene flow has ceased.

Another feature of species clusters which causes puzzlement is parallelism. Sympatric species frequently show similar adaptations yet have obviously come from quite separate ancestry. This result is not surprising if the two parallel entities have basically similar genetic systems and adaptations. What natural selection can do with one gene pool it can do with another, although the results will never be identical and coexistence is frequently possible.

Oceanic islands have long served as natural laboratories for the observation of evolutionary phenomena. Genetic data are now being gathered on the extraordinary drosophilid fauna of Hawaii. More than 400 endemic species, deriving ultimately from very few ancestors, have been described (Hardy, 1965). Preliminary evidence indicates that most species are endemic to single islands yet the species groups are widespread, suggesting miltiple proliferations.

A most striking feature is the great karyotypic stability among clusters of closely related species (Carson, Clayton, and Stalker, 1967). In a number of instances, groups of homosequential species occur. In such species the metaphase and gene order karyotype is identical, not having been altered by inversions or translocations. Although morphologically and ethologically diverse, such species are often found sympatrically and do not seem to represent strikingly different adaptive types.

There is no apparent reason why speciation should differ in its basic character between a continent and an oceanic archipelago. In the case of islands, especially high oceanic ones, the essential difference is evidently the great role played by isolation due to the physical forces of the environment. The role played by possible past population flushes can only be conjectural at this stage, although there is considerable evidence, from study of introductions in modern times, that great flushes may occur (see, for example, Wodzicki, 1965).

CONCLUSION AND SUMMARY

The thesis is set forth that extraordinary population flushes, mediated primarily by environmental factors, occur at rare intervals in most species. In species which have a high recombination index and carry much genetic variability, the flush will result in the release of enormous genetic variability as the population size reaches its crest. Although lethals and other drastics will be eliminated by selection, the flush is characterized by a breakdown in the genetic basis of fitness. A myriad of gene combinations appears; many of these would be mildly dysgenic under ordinary circumstances. Following widespread dispersal, the population may crash or at least be greatly reduced from the crest of the flush. This process will result in the formation of many isolated demes some of which may represent new colonizations for the species. Selection quickly eliminates non-breeding individuals and then proceeds to the more significant aspect, namely, the elimination of demes with relatively unfit gene pools. Those that are not eliminated may indeed be very few in number but are nevertheless of great significance. It is considered likely that there will be few, if any, new adaptations arising at this phase; interdeme selection is considered as primarily a negative process. The new isolates are forced to inbreed in small populations, and natural selection may first favor reciprocal adjustments of the sexes in reproduction. This, together with the well-known profound effects of random drift on the gene frequencies in the new deme, may result in reproductive isolation from the parent group should another flush occur. Thus, a new species is born under circumstances where adaptation is not the guiding force. As time goes on, intra-species phenomena of mutation, recombination and selection may forge new adaptations, but when newly formed, the species will not be adaptively very different from the parent. This idea appears to explain some of the puzzling features of sympatry, sibling species, species clusters, "non-adaptive" characters, and parallelism.

LITERATURE CITED

Allee, W. C., *et al.*, 1950, *Principles of Animal Ecology,* Philadelphia, Saunders Co., pp. 837

Andrewartha, H. G., and L. C. Birch, 1954, *The Distribution and Abundance of Animals,* Chicago, Univ. Chicago Press, pp. 782.

Band, H. T., 1964, "Genetic Structure of Populations. III. Natural Selection and Concealed Genetic Variability in a Natural Population of *Drosophila melanogaster,*" *Evolution,* 18:384–404.

Brncic, D., 1954, "Heterosis and the Integration of the Genotype in Geographic Populations of *Drosophila pseudoobscura,*" *Genetics,* 39:77–88.

Buri, P., 1956, "Gene Frequency in Small Populations of Mutant Drosophila," *Evolution,* 10:367–402.

Butler, L., 1953, "The Nature of Cycles in Populations of Canadian Mammals," *Canadian J. Zool.,* 31:242–262.

Cannon, G. B., 1963, "The Effects of Heterozygosity and Recombination on the Relative Fitness of Experimental Populations of *Drosophila melanogaster,*" *Genetics,* 48:919–942.

Carson, H. L., 1957, "The Species as a Field for Gene Recombination," E. Mayr, Ed., *A.A.A.S. Publ.* No. 50. pp. 23–38.

———, 1961, "Heterosis and Fitness in Experimental Populations of *Drosophila melanogaster,*" *Evolution,* 15:495–509.

Carson, H. L., F. E. Clayton and H. D. Stalker, 1967, "Karyotypic Stability and Speciation in Hawaiian Drosophila," *Proc. Nat. Acad. Sci.* 57:1280–1285.

Chitty, D., 1957, "Self-regulations of Numbers Through Changes in Viability," Cold Spring Harb. Symp. Quant. Biol., 22:277–280.

Clough, G. C., 1965. "Lemmings and Population Problems," *Amer. Sci.,* 53:199–212.

Dobzhansky, Th., 1948, "Genetics of Natural Populations. XVI. Altitudinal and Seasonal Changes Produced by Natural Selection in Certain Populations of *Drosophila pseudoobscura* and *Drosophila persimilis.*" *Genetics,* 33:158–176.

———, 1963, "Genetics of Natural Populations. XXXIII. A Progress Report of Genetic Changes of *Drosophila pseudoobscura* and *Drosophila persimilis* in a Locality in California," *Evolution,* 17:333–339.

Dubinin, N. P., and G. G. Tiniakov, 1945, "Seasonal Cycles and the Concentration of Inversions in Populations of *Drosophila funebris,*" *Amer. Natur.,* 79:570–572.

Elton, C., 1942, *Voles, Mice and Lemmings, Problems in Population Dynamics,* Oxford Univ. Press.

———, 1958, *The Ecology of Invasions by Animals and Plants,* London, Methuen & Co.

Epling, C., D. F. Mitchell, and R. H. T. Mattoni, 1957, "The Relation of an Inversion System to Recombination in Wild Populations," *Evolution,* 11:225–247.

Errington, P. L., 1957, "Of Population Cycles and Unknowns," Cold Spr. Harb. Symp. Quant. Biol., 22:287–300.

Falconer, D. S., 1960, *Introduction to Quantitative Genetics*, New York, The Ronald Press, pp. 365.

Ford E. B., 1964, *Ecological Genetics,* London, Methuen & Co., pp. 335.

Ford, H. D., and Ford, E. B. 1930, "Fluctuation in Numbers and Its Influence on Variation in *Melitaea aurinia,*" *Trans. Roy. Ent. Soc. Lond.,* 78:345–351.

Gause, G. F., 1934, *The Struggle for Existence,* Baltimore, Williams and Wilkins, pp. 163.

Haldane, J. B. S., 1957, "The Cost of Natural Selection," *J. Genet.,* 55:511–524.

Hardy, D. E., 1965, Diptera: Cyclorrhapha. II. *Insects of Hawaii,* Vol. 12, Honolulu, Univ. of Hawaii Press, pp. 814.

Hewitt, O. H., 1954, "A Symposium on Cycles in Animal Populations." *J. Wildl. Mgmt.,* 18:1–112.

Kluijver, H. N., 1951, "The Population Ecology of the Great Tit, *Parus m. major L.,*" *Ardea,* 39:1–135.

Lack, D., 1954, *The Natural Regulation of Animal Numbers,* London, Oxford Univ. Press, pp. 343.

Levitan, M., 1951, "Response of the Chromosome Variability in *D. robusta* to Seasonal Factors in a Southwest Virginia Woods," *Genetics,* 36:561–562.

Lewontin, R. C., 1965, Selection for Colonizing Ability, In *The Genetics of Colonizing Species,* H. G. Baker and G. L. Stebbins, eds., New York, The Academic Press, pp. 79–91.

Mac Arthur, Robert H., 1955, "The Causes of Fluctuations in Animal Populations and a Measure of Community Stability," *Ecology,* 36:533–536.

Mayr, E., 1954, Change of Genetic Environment and Evolution. In *Evolution as a Process,* J. Huxley, Ed., London, Allen and Unwin, pp. 157–180.

Oshima, C., and O. Kitagawa, 1963, "The Persistence of Deleterious Genes in Natural populations of *Drosophila melanogaster.* III. Difference Between Lethal and Semilethal Genes." *Proc. Jap. Acad.,* 39:125–130.

Stebbins, G. L., 1950, *Variation and evolution in plants,* New York, Columbia Univ. Press, pp. 643.

——, 1959, "The Role of Hybridization in Evolution," *Proc. Am. Phil. Soc.,* 103:231–251.

Vetukhiv, M., 1954, "Integration of the Genotype in Local Populations of Three Species of *Drosophila,*" *Evolution,* 7:241–257.

Williams, G. C., 1966, *Adaptation and Natural Selection,* Princeton, Princeton Univ. Press, pp. 307.

Wilson, E. O., and Brown, Jr., W. L. 1958, "Recent Changes in the Introduced Population of the Fire Ant *Solenopsis saevissima (Fr. Smith).*" *Evolution,* 12:211–218.

Wodzicki, K., 1965, *The Status of some Exotic Vertebrates in the Ecology of New Zealand,* In *The Genetics of Colonizing Species,* H. G. Baker and G. L. Stebbins, eds., New York, The Academic Press, pp. 425–460.

Wright, S., 1964, Stochastic Processes in Evolution. In *Stochastic Models in Medicine and Biology,* J. Gurland, ed., Madison, Univ. of Wisconsin Press, pp. 199–244.

-10-

The Regulation of Numbers and Mass in Plant Populations

JOHN L. HARPER
School of Plant Biology
University College of North Wales, Bangor

By far the greater bulk of population theory, experiments, and practice has been developed from observations on animal populations. Any study of the population biology of plants runs the risk, therefore, of uncritically accepting a population theory which is not derived for organisms with essentially plant-like properties. It therefore seems worthwhile to make some attempt to define the essential qualities of plants that affect their behavior in populations.

There are three major ways in which the biology of populations is studied and interpreted.

Spatial and Temporal Distribution

This normally takes the form of detailed mapping, on a geographical or ecological scale, of the distribution of particular species and the correlation of this distribution with physical features of the environment. It may be extended into formal classification such as that of continental phytogeographers (e.g., Braun-Blanquet, 1951), hierarchical, computer-based systems of classification such as those of Williams and Lambert (1959), or systems of ordination (Curtis and McIntosh, 1951; Greig-Smith, Austin, and Whitmore, 1967). This is characteristically a botanist's approach, because the fixity of individuals makes for easy mapping.

Energetic Studies

A population may be defined and described in its relationship to the ecosystem in which it is found. This may involve studies of energy flow or of nutrient cycling, and focuses attention on the diversity of ways in which different species utilize natural resources. The niche, essentially an animal ecologist's concept, defines the position of a species or population within a community by reference to the position in a food chain, the food that it eats and the food that it in turn provides to the organisms that eat it.

Population Dynamics

The population may be studied in terms of its numbers and their rates of change. This is again more usually an animal ecologist's approach because many small animals are especially amenable to population study in the laboratory.

Although these three approaches to the study of populations have become increasingly well defined, and specialized, the aspects of population biology that they describe are strongly interrelated, perhaps more obviously so in the case of plant populations than with animals.

THE POSITION OF PLANTS IN FOOD CHAINS

Entry of Energy Into the First Trophic Level

The position of the plant population in food chains is critically defined by the characteristics of the two chlorophyll molecules enabling the fixation of incident radiation to occur in the limited range of 4,000–7,000 Angstroms. Higher plants, with the exception only of parasites and chlorophyll-less saprophytes, lack any choice in energy supply. This immediately contrasts the natural mixed plant population in an ecosystem with the animal population in which diversity of diet and the existence of species in diverse positions in a complex food web permits a high degree of specialization in energy resource and hence a high degree of diversification associated with the variety of foods. If we were to try to devise a plant population which partitioned energy resources in as diverse a fashion as do animals, we would presumably have to imagine plants with different pigment systems capable of using light at different wavelengths; in fact, we find a virtual omnipresence of the same energy-fixing system in all higher plants. This system is itself rather narrowly restricted, leaving a large part of the theoretical "energetic niche" unoccupied. Although photosynthetic systems exist (for example, amongst bacteria) which will fix radiant energy up to and beyond 8,000 Ångstroms, higher plants are restricted to fixation in the 4,000–7,000 Ångstrom region, leaving some 56 per cent of incident radiation as an untapped, unexploited resource.

Because populations of higher plants derive their energy directly from solar radiation, the behavior of the population is fundamentally affected by the intensity and the rhythm of the radiation. In a given locality, both the annual cycle and the total receipt of solar energy vary within very narrow limits. The solar radiation falling on an area of land is fixed, therefore, with some precision, and its availability to the vegetation as a whole is not affected by the process of consumption (although mutual shading will determine the way different individuals share the limited incoming energy). It is, needless to say, impossible to generate a predator/prey cycle in which the plant is a predator and solar energy is the prey. The source of all animal food is capable of exponential increase, that of plants is not.

The "consumable" requirements for growth of a population of higher plants are light, water, mineral nutrients, and carbon dioxide. The position of water as a consumable is somewhat odd because only very limited quantities are fixed in growth, but its continued passage through the plant and loss to the atmosphere is a necessary consequence of the apparatus used for assimilating carbon dioxide. The stomatal mechanism which permits carbon dioxide to enter the leaf permits water to escape. The availability of water, therefore, determines the ability of a plant to utilize solar radiation. Nutrient consumption, in contrast, involves accumulation within the plant body, and the effective working of particular physiologic systems depends on the particular combinations of nutrients available to the plant. A specific nutrient requirement such as that of molybdenum for the legume/bacterial association, specific ability to tolerate low levels of a normal nutrient, such as iron in calcareous soils, ability to tolerate the peculiar nutrient combinations present in soils of different pH, and the ability to tolerate the presence of toxic minerals (e.g., selenium, copper, zinc, aluminium), are all qualities in which plant species or populations may differ from one another.

Resources for the growth of a plant population are not discrete except at the atomic or quantal level. There are no unit bodies of food or prey to be searched for or captured. The growth of a population is, therefore, under limitations that are spatially imposed by the light falling on a unit area, the water available for transpiration in a unit area, and the nutrients available in the soil in a given area.

The limited resources of light energy falling on an area of land are trapped by a leaf canopy. Agronomists have been at some pains to determine the parameters of a crop population which determines the rate at which it fixes energy. Not surprisingly, it turns out that the number of plants present is an inefficient indicator of fixation and that the actual area of leaf exposed by the population is more efficient. The parameter Leaf Area Index (Watson, 1952; Brougham, 1960) has been employed as a useful measure of the area of leaf exposed per unit area of land. Integrated over time as the Leaf Area Duration, it correlates well with the rate of growth of dry matter. Depending on the density of leaves, the angle at which they are held, and the direction of incident radiation, a population will exert a degree of mutual shading. A Leaf Area Index may develop which is so high that the lower leaves in a population receive insufficient light for photosynthesis to occur as fast as respiration. Indeed, dense populations of plants may be made into more effective energy-fixing systems by deliberate leaf removal or plant thinning to bring the populations as a whole to an "optimal" Leaf Area Index. Figure 1 illustrates the way in which the

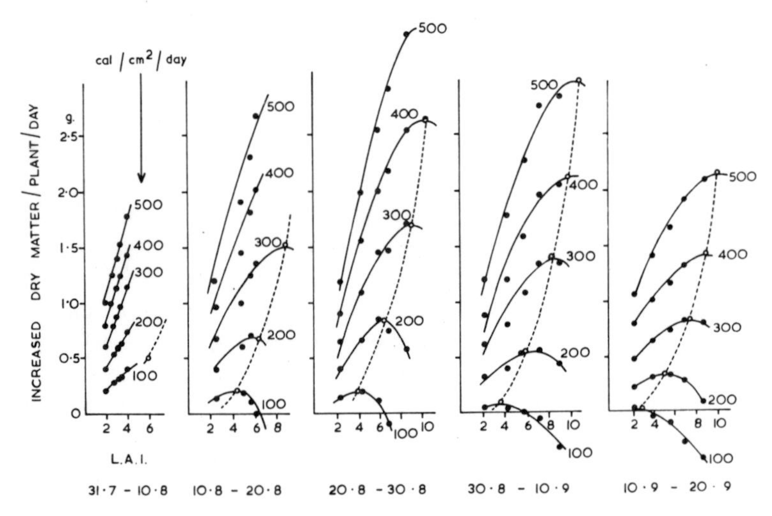

FIG. 1. The relationship between plant growth and Leaf Area Index at five different intensities of incident radiation and at five different growth stages. The "optimal" L.A.I. is indicated by the broken line.

The five stages of growth are represented by the five graphs, reading from left to right: maximum tillering stage, young panicle developing, "booting" to heading, heading to early ripening, vigorously ripening (redrawn from Takeda, 1961).

density of the leaf population determines the rate of dry matter fixation at various light intensities and stages of growth in rice. Because of the high vegetative plasticity of plant growth, the same Leaf Area Index may quickly be developed from a very wide range of densities of plants sown. The individual plants quickly become integrated into an energy-trapping canopy in which the performance of individual plants has been dominated by the characteristics of the population (Harper, 1964b). Because the loss of water from the plant is itself determined by the radiant and advective energy received by the unit area of land, the water relations of a population also tend to dominate the water relations of the individual.

Diversification in the use of incident radiation by a canopy may be seen in the varied angles at which leaves of different species are inclined, fitting them for optimal light interception at different incident intensities and directions (see discussion by De Wit, 1965). Perhaps more strikingly, non-synchronous exploitation of incident radiation can often be detected amongst the species in a community. This is particularly strongly developed in the pre-vernal, vernal, and aestival phases of leaf production by the species of British woodland (Tansley, 1949). Even closely related species living in the same area may have subtly different cycles of leaf production:

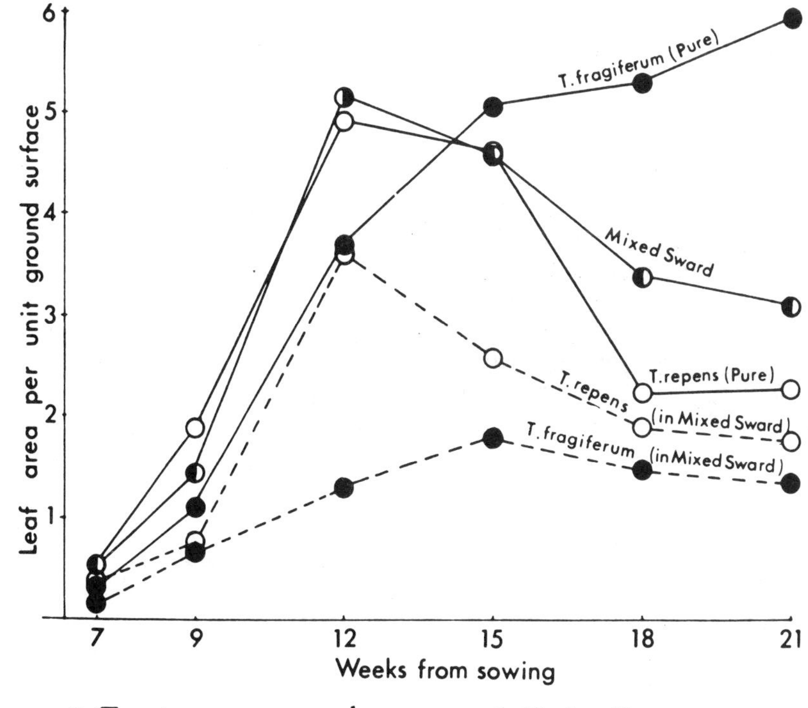

———○ *T. repens* pure sward ———● *T. fragiferum* pure sward
———◐ Mixed sward - - - ○ *T. repens* in mixed sward
- - - -● *T. fragiferum* in mixed sward

Figure 2. The development of leaf area in single species populations of *Trifolium repens* and *T. fragiferum* and in mixtures of the two. Each species in pure stand was grown at a density of 70 plants per 1000 cm². In mixtures *each* species was at a density of 35 plants per 1000 cm² (From Harper and Clatworthy, 1963).

Figure 2 illustrates such non-synchrony for two species of *Trifolium, T. repens* and *T. fragiferum.*

Where the growth cycles of individuals in a population are synchronous and light is nearly fully intercepted, victory goes to the individuals which achieve height quickest, whether by producing taller leaves (as in the grasses), longer petioles, taller stems, or more perennial stems. The studies of Black (1960) using inter- and intravarietal mixtures of forms of subterranean clover, *T. subterraneum*, demonstrate very elegantly the overriding importance of a high canopy for survival in a struggle for existence. The advantage of a high canopy is gained, however, by expenditure of a greater proportion of the plant's resources on the non- or weakly photosynthetic organs supporting this high canopy.

Passage of Energy to the Second Trophic Level

Plant populations are peculiar in the way that they pass energy to the next stage in the food chain. Most predator/prey interactions in the animal kingdom involve the consumption of individuals by others. A whole animal is eaten, or at any rate killed, by the predator. There is, of course, some predation on plant populations that involves death of the individual, for example, the eating of seedlings by slugs and snails, or the lethal activities of some stem and root boring insects. However, the major contribution of plants to the food of herbivores occurs by the removal of only parts of the plant system of which the remainder proceeds to regenerate. The best known example of this is the management of a cattle or sheep population on grassland in which the herbivore population is adjusted in efficient management practice to consume the interest on, but not the capital of, the grass plants. It is difficult to visualize any comparable system operating in animal populations feeding on animals, the closest approach presumably being classic cases of parasitism rather than anything traditionally called a predator/prey relationship. An obvious analogue would be a predator which removed only one arm from a starfish, allowing the starfish to regenerate its arm before another was consumed. However, animal populations seem not to have evolved feeding habits of this nature except in rare instances like the predation of colonial hydroids by molluscs. At the herbivore end of the plant's position in the food chain, predator/prey cycles can, of course, be generated if more than the capital of the grazed plant is taken. Modern agricultural management of grassland that involves rotational grazing with periods of rest for the grassland while the livestock are moved from paddock to paddock is an example of a man-managed predator/prey oscillation.

One of the most peculiar properties of plant populations in an ecosystem is that the greater part of the energy fixed in plant material is unavailable to herbivores because it is in the form of cellulose or lignin. The highly restricted distribution of enzyme systems capable of dealing with the structural components of the plant is a great mystery. Cellulases are restricted to bacteria and some fungi, slugs and snails, and some longhorn bettles. Only by the employment of bacterial symbionts does plant cellulose become available as an energy source for higher animals; the remainder finds its way to the decomposer industry (Macfadyen, 1964).

Predation by herbivores may act, of course, as a powerful selective force in determining the species composition of vegetation or the genotypic composition of a population. Charles (1961) demonstrated profound shifts in the composition of populations of *Lolium perenne* and other grasses following periods of intensive cattle grazing, and Ehrlich and Raven (1965) have suggested that the herbivore/plant relationship is a source of radiative evolution as exemplified in the Papilionidae. The way in which plant population size may be regulated by herbivores proves very difficult to unravel. *Hypericum perforatum*, a weed introduced into California, spread rapidly, forming dense stands in open grassland and in woodland. The introduction of the biological control agent, *Chrysolina quadrigemina*, produced striking changes in the distribution. The weed changed from being very abundant in grassland to being represented by only isolated individuals—the population density perhaps being regulated by the limited area of search of the beetle. However, populations in woodlands remained almost unchanged because *Chrysolina* does not readily lay eggs in the shade (Huffaker and Messenger, 1964). It is difficult to envisage how a newcomer to the Californian scene, finding the present distribution and abundance of *Hypericum*, would ever design an experiment that would reveal that *Chrysolina* was both regulating density and determining this rather peculiar restricted distribution. He would probably search for evidence of shade tolerance in *Hypericum* to explain the distribution!

LEVELS OF POPULATION BEHAVIOR

The Individual as a Population

In its growth, an individual higher plant behaves essentially like a population. It is a tenet of classical morphology that a plant can be regarded as a repeating system of which the unit is a leaf on a segment of stem with an axillary bud. This basic unit of growth repeats to form the complex branching structure of a tree, or, spreading laterally as well as vertically, gives the vegetative spreading habit of a strawberry or a grass. This "population" of shoots may be connected to a single root system conferring an identity on the population as an individual. Where there is lateral vegetative spread, independent root systems may become established and the original genetic individual may lose its identity and even its physical continuity. Harberd (1961, 1962, 1963) has shown that clones of *Festuca rubra, Festuca ovina*, and *Trifolium repens* derived initially from a single seedling may extend over considerable areas; a single genotype of *Festuca rubra* was found over an area 240 yards in diameter. However, even the more "orthodox" plant, growing and branching vertically, has essential properties of a population with a capacity for exponential increase in the numbers of its basic units. This capacity becomes more and more limited as the increasing number of units interfere with each other by putting increasing demands on limiting resources. Individual units of the plant shade each other or provide demands for water and nutrient in excess of the root system's supplying capacity. Thus the individual palnt, even when grown in isolation, may show the essential phases of a logistic growth curve (Ashby, 1936).

The development of an individual plant as a population is a manifestation of its plasticity. The individual may find expression in a wide variety of forms depending on the stresses and supplies available from the environment; the number of leaves, branches, flowers, fruits, and seeds is an expression of how far the individual plant has developed as a population. Where vegetative reproduction occurs, a lateral shoot or ramet may start life provided with food supplies translocated from the parent shoot (Sagar and Marshall, 1966), then gain independence, and then by virtue of proximity to the parent shoot interfere with its supplies of light, water, and nutrients in the same way as a neighboring plant established from an independent seed. The resemblance of this model to budding reproduction in Hydra is only superficial because the Hydra maintains the capacity for daughters and parent to wander away from each other. In plants, the distance at which a daughter ramet establishes from the parent is a species-specific function of the type of rhizome, stolon or runner that is developed. The adjustment is sensitive to environmental regulation, but rather little is known about the way in which the spatial positioning of the offspring of vegetative reproduction is determined by environmental conditions.

Populations of Individuals

The establishment of plant populations from seed is a much more disorganized affair than the process of vegetative reproduction. Seed density, like pollen density, tends to fall off exponentially with distance from the parent, and (with certain very insignificant exceptions such as the burying movement of the awns of certain plant propagules) there are no processes equivalent to search, migration, or settlement which contribute to adjusting densities in animal populations. The propagule of the higher plant represents a high element of maternal care, the zygote developing to an advanced stage before being arrested in development and dispersed in a state of suspended animation or dormancy.

Species differ strikingly in the specific conditions required to break seed dormancy. The size of a plant population established from seed is a function of the availability of seed and the frequency of sites (where by the word "site" is meant something of much the same size as that of the seed itself) which provide the specific requirements for breaking dormancy (Harper, Williams, and Sagar, 1965). If both seeds and sites are abundant, density dependent mortality of seedlings may follow (Harper and McNaughton, 1962), and this process of thinning may continue as the growing plants increase their demands and interfere with each other more and more strongly. This process has been studied in *Erigeron canadensis* by Yoda *et al.* (1963), who showed that there was a mutual adjustment between the growth of individuals and the rate of loss from the population during the annual cycle of growth in pure populations. Moreover, as conditions of soil fertility were improved by addition of fertilizer, individuals grew faster and the population self-thinned to a greater extent. The two processes of population regulation by change in numbers and by change in individual mass are closely interrelated.

An important consequence of the plastic growth habit of plants is that production per unit area tends to become independent of the number of plants over a wide range of densities. The almost perfect compensation of individual plant weight for changes in density is illustrated in Figure 3 for *Trifolium repens* and *T. fragiferum* under three contrasted regimes of water supply. When the water table was maintained high, establishment of seedlings was excellent, but their growth was so restricted that even at the highest density the plants did not interfere with each other sufficiently to produce detectable change in individual dry weight. In the more effectively drained regimes, growth was more rapid, and the stress of density is reflected in a compensating decline in individual plant weight. Most of the information that we have about such yield/density relationships have been derived from agricultural crops or weeds (see, e.g., Donald, 1963).

SUSPENDED ANIMATION

Populations of higher plants may remain dormant for many years. The dormant propagule is normally the seed, and it represents, therefore, an

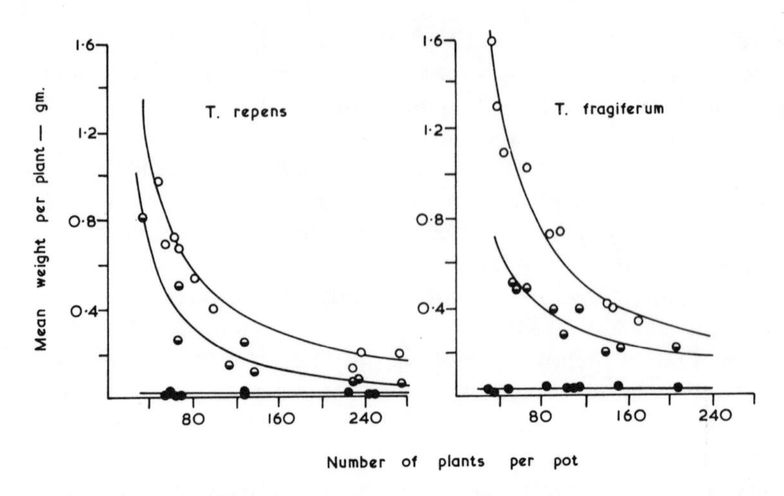

Figure 3. The relationship between mean plant dry weight of *Trifolium repens* and *T. fragiferum* and plant density under three contrasted water regimes. ○ pots freely drained; ◒ water table 17.5 cm below soil surface; ● water table within 1 cm of soil surface (from Clatworthy, 1960).

arrested stage in the individual's development. Dormancy may function like diapause in insects to determine strictly seasonal population behavior. In many cases, however, germination is prevented for indeterminate periods because of the absence of some specific environmental requirement for germination, such as light or oxygen or the presence of some inhibiting agent in the environment such as carbon dioxide. Buried viable seed populations may reach fantastically high levels in agricultural land, but there appear to have been no studies of long-established natural vegetation except of permanent grassland. Champness and Morris (1948) found values up to 5 × 10⁶ seeds per acre of *Stellaria* media in pastures and 10.8 × 10⁶ seeds per acre in arable land, and 12.6 × 10⁶ seeds per acre of *Poa annua* in grassland soils, 25.5 × 10⁶ in arable soils. Seed of *Juncus* spp. has been reported at densities of 60 × 10⁶ per acre in upland grassland soils. According to Roberts (1962), the buried weed seeds in horticultural land show a characteristic decay rate with constant half life. The buried seed populations, therefore, represent the results of continuing recruitment and continuing loss—a sort of "running mean" from the past populations of seed bearing plants. I have discussed elsewhere the role that this dormant population may have in buffering the species against short-term selectional change (Harper, 1956). The phenomenon has important ecological consequences for it ensures a reservoir of propagules of a species in an area ready to exploit a favorable environment that may suddenly be created. In many grassland habitats we have evidence (Cavers and Harper, 1967; Putwain and Harper, 1968) that establishment of new individuals from seed in a mixed grassland population depends on local disturbance such as wormcasts, molehills, footprints, etc. It is from the buried viable weed seed population that recruits to the disturbed areas most often come. A similar, though not nearly so long-lived, phase has been reported by Chippindale (1948) for grasses and by Cavers and Harper (1967) for *Rumex* species where it has been observed that seedlings may persist through periods of a year or more, losing weight very steadily but remaining alive and capable of rapid growth if the immediate environment becomes mollified.

EXPERIMENTAL STUDIES ON PLANT POPULATIONS

Much of the population ecology of plants bears the very strong imprint of the peculiar properties of plants as experimental material. Seed provides easy material for setting up density experiments, and because of the obvious agricultural implications of minimizing seed rates and assuring optimal yields, a great many studies have been made of the density reactions of crops. It is relatively easy to grow mixtures of species, particularly annuals, in order to study the effects of proportion and of density. Most such experiments have involved the synthesis of communities of single or two-species mixtures, and attention has been focussed on (a) the timing and specificity of requirements for germination, (b) the sieve-like behavior of the heterogeneous soil environment which selects the types and numbers of successful individuals, and (c) the examination of community diversity and its explanation in terms of the differential behavior of species to each other's density in experimental models (Harper, 1967).

A Model for the Study of Density and Proportions in a Mixture of Two Plant Forms

An example of the latter type of experimental model is illustrated in Figure 4, the results of an experiment made on mixtures of oil seed and fiber flax (both varieties of *Linum usitatissimum*). Such a mixture never occurs in nature or in agricultural practice, but was chosen as a result of a previous experiment in which six varieties of oil and fiber flax had been grown in all possible combinations to determine their ecological combining ability (Harper, 1964a). This preliminary experiment had shown that the total dry matter production of certain mixtures of oil and fiber flax could

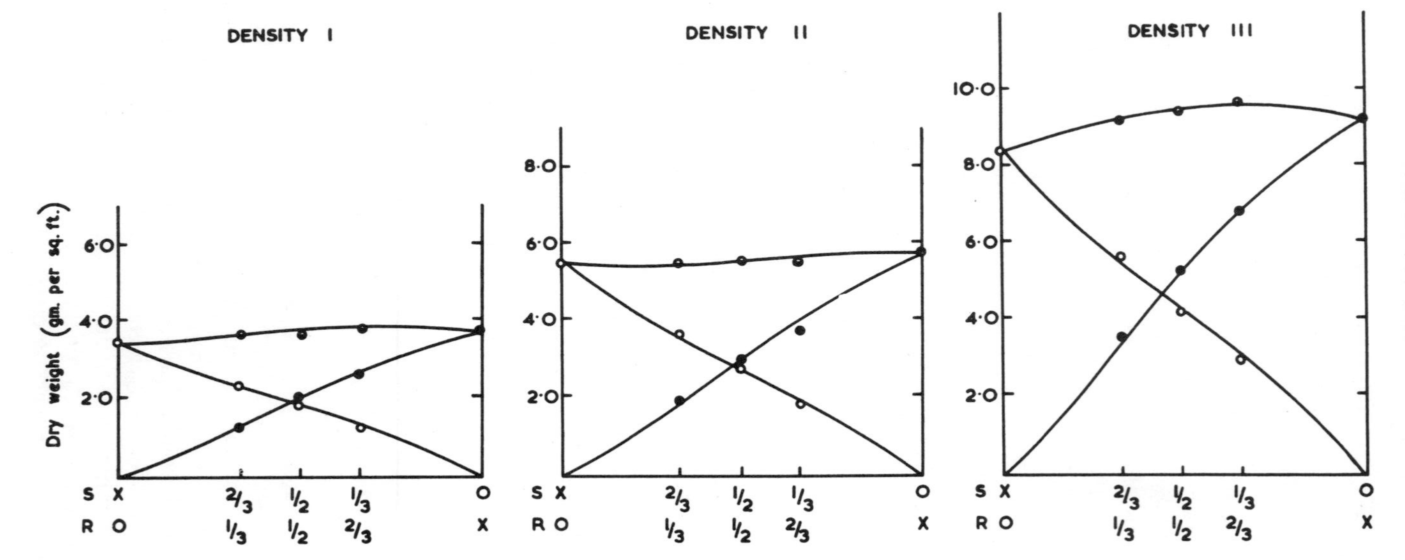

Figure 4(a). The influence of time, density, and proportions on dry matter production by oil seed and fiber flax *(Linum usitatissimum)* in mixtures. Dry weight 21 days after seedling emergence. S = Stormont Gossamer (fiber flax) ●; R = Redwing (oil seed flax) ◒; Combined yield of mixtures ○. X, $^2/_3$, $^1/_2$, $^1/_3$, 0 indicate proportions of the two varieties in mixtures. Density: I = 800 seeds per 1000 cm^2; II = 1600 seeds per 1000 cm^2; III = 3200 seeds per 1000 cm^2.

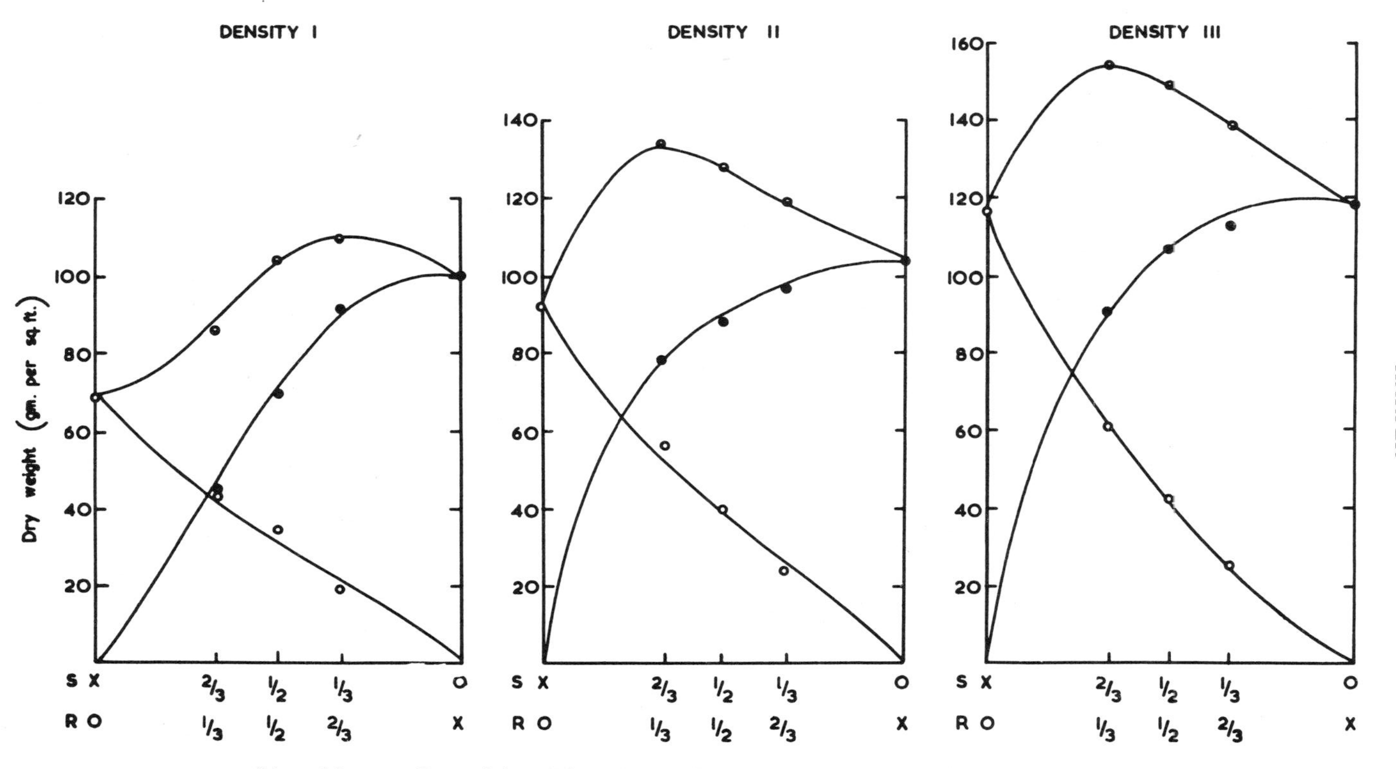

(b). as (a) except: Dry weight 119 days after seedling emergence (from Obeid and Harper, in press).

exceed that of the highest yielding component grown in pure stand. Seed of the two cultivars was sown at the same over-all density but at varying proportions including pure stands of each cultivar. This form of experimental design makes it possible to compare the reaction of a plant to the presence of others of its own sort with its reaction to plants of another variety, thus contrasting in one experimental design, intra- and intervarietal effects. In Figure 4a, it can be seen that, in respect to dry matter production, early in the growing season the cultivars each reacted to the other as they reacted to themselves (that is to say inter- and intravarietal effects were similar). As a consequence, the dry matter production from the variously constituted mixtures lies in the range predictable from the yield of the pure stands. In contrast, at a later stage in development, Figure 4b, mixtures produced a greater yield of dry matter than either of the pure stands. In this later growth phase, the reaction of each cultivar to its own density was different from its reaction to the other variety. As a result, the two forms in combination more effectively used environmental resources to the end of dry matter production. A detailed analysis of the growth cycles of the two cultivars (Obeid, 1965) showed that they were not completely synchronous. Fiber flax completed its growth cycle earlier than oil seed flax. In mixtures, the fiber flax ripened and its leaves withered at a time when the oil seed flax was still capable of making further growth. Oil seed flax in mixtures increased in dry weight, more than in a pure stand, with the result that the two cultivars in mixture avoided the full intensity of synchronous density stress.

It is important to recognize that the superiority of such a mixture compared with a pure stand need not imply any balanced mutualism or facilitation. The superior production by mixed stands can be accounted for wholly in terms of the reduced interference each cultivar offered to the other.

The Regulation of Density in "Natural" Populations of Rumex acetosella

An alternative to the synthetic population experiments exemplified by the previous example is the analytic approach in which natural or seminatural populations of plants may be altered in an attempt to expose the underlying causes of regulation of dry matter production or of numbers. Putwain, Machin, and Harper (1968) sowed seed of *Rumex acetosella* at six densities into a grassland sward which had received three contrasting treatments. The sward contained twelve species of higher plants dominated by grasses. *Rumex acetosella* was an abundant constituent of the selected grazed sward, which had established as a result of the natural colonization of arable land over a period of four years. Seed was sown on plots in which natural seed dispersal of *Rumex acetosella* was also permitted, and also on plots where natural seed dispersal was prevented by removing the female inflorescences as they appeared or (because this treatment might produce abnormalities) bagging the inflorescences to prevent seed dispersal. The size of the population was followed by counting vegetative shoots and seedlings at intervals over a period of a year. *Rumex acetosella* reproduces vigorously from root buds, and a major aim of this experiment was to determine how far regulation of the populations of this species operated through seed or vegetative reproduction. The data on population size was analyzed by selective multiple regression analysis in an attempt to relate the population size at any given time to its size on previous sampling occasions. In the environment of the experimental sward, very few seedlings of *Rumex acetosella* became established. Contributions to the population from seed were insignificant compared with those from vegetative reproduction. Throughout the spring and summer, a positive linear function of the initial number of mature vegetative units completely accounted for the total size of the population. These linear regression constants increased in value from April to June and then declined. There was, however, tentative evidence of density dependent regulation both at mid-summer and in the early fall when negative quadratic functions of the initial population approached significance. Where self-seeding was prevented, either by bagging or removal, the behavior of the population of mature individuals was the same as that of the control plots, suggesting that preventing a plant from setting seed did not affect its vegetative performance. This type of experiment is not without its difficulties. Once cotyledons have been lost, it is very difficult to distinguish a seedling from a young vegetative shoot: it is quite impossible, of course, to distinguish individual shoots arising as a clone from individuals which belong to different clones. The experiment also neglected any changes in components of the sward other than *Rumex acetosella.* The role of other species in such a mixed population may be determined by removing them and observing the effects on *R. acetosella.*

The Role of Associated Species in the Regulation of Populations of Rumex acetosella *and* R. acetosa *in Grassland*

A series of experiments was made by P. D. Putwain to determine the role of associated species in determining the density of *Rumex acetosa* and *Rumex acetosella* in grassland. Two areas of hill grassland were chosen for these studies, the first containing abundant *R. acetosa,* together with a characteristic upland grassland flora of grasses such as *Holcus lanatus, Festuca rubra, Festuca ovina,* and *Agrostis tenuis,* and a range of dicotyledonous broad-leaved species such as *Potentilla erecta* and *Galium saxatile.* The second site, in which *Rumex acetosella* was abundant, was a sward composed largely of *Festuca ovina* with *Galium saxatile* and a number of other frequent broad- leaved species. Various components of the flora were selectively removed by applying specific herbicide treatments. The sodium salt of 2,2-dichloro-propionic acid (Dalapon) was applied to remove all grass species. Spot treatments of selected species with 2,4-dichlorophenoxyacetic acid and Tordon 22K (a potassium salt of 4-amino-3,5,6-trichloropicolinic acid) removed all non-gramineous species and left the sorrels and grasses. Some plots were sprayed with 1,1-dimethyl-4,4-bipyridylium dichloride (Paraquat) to destroy all vegetation except *Rumex acetosa* and *Rumex acetosella.* This treatment produced some damage to the *Rumex* spp., but they quickly regenerated, unlike the other species. A further treatment involved spot application of Tordon to plants of *R. acetosa* and *R. acetosella,* thus removing the existing population of these species but no others. In general, good selectivity was obtained by these treatments. At each site seed was sown. At the first (*Rumex acetosa*) site, densities of 1, 75, and 750 seeds were sown per 30 cm + 30 cm square. At the second (*Rumex acetosella*) site, seed was sown at densities of 1, 150, and 1,500 seeds per 30 cm square. At the first (*Rumex acetosa*) site, when grasses were removed from the habitat the population of shoots of *Rumex acetosa* increased rapidly. However, when the non-gramineous species were removed, there was no effect on the population of Rumex acetosa (Figure 5). These results have been expressed diagrammatically as a niche model in Figure 6a. A peculiarity in the response of *Rumex acetosa* to removal of its associates was that its population increased less if both dicotyledons and grasses were removed than if grasses alone were removed. This may mean that in some way the presence of the dicotyledonous members of the community extends the niche of *Rumex acetosa.* This feature is indicated in Figure 6a as a semi-circle extending the niche of *R. acetosa.* However, it may be that the greater sensitivity of *R. acetosa* to Paraquat than to Dalapon is responsible for this observation.

Seedling establishment did not show the same reaction to removal of existing vegetation as did vegetative expansion. Successful seedling establishment is indicated in Figure 6 by small cross-hatched circles. Seedlings of *Rumex acetosa* were established from sown seed when the grasses were removed or when the dicotyledonous species were removed but not when *Rumex acetosa* alone was removed from the community. This is further evidence that the fundamental niche of the seedling stage of a plant is different from that of the adult stage. Seed of *Rumex acetosa*

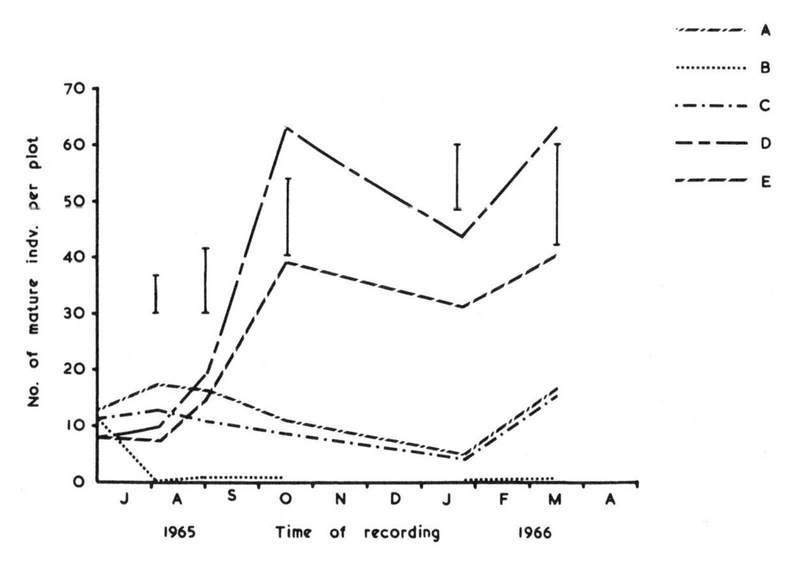

Figure 5. The influence of the selective removal of various components from a mixed grassland sward on the numbers of vegetative shoots of *Rumex acetosa*. Selective removal was by appropriate herbicide application (see text) made on June 18, 1965. A = control—untreated; B = removal of *Rumex acetosa* (Tordon spot treatment; C = removal of dicotyledonous fraction of flora (except R. *acetosa*) (Tordon and 2,4–D spot treatment); D = removal of grasses (Dalapon treatment); E = removal of all vegetation except *R. acetosa* (Paraquat treatment). Fiducial limits are indicated at P = 0.05 (Putwain and Harper, in preparation).

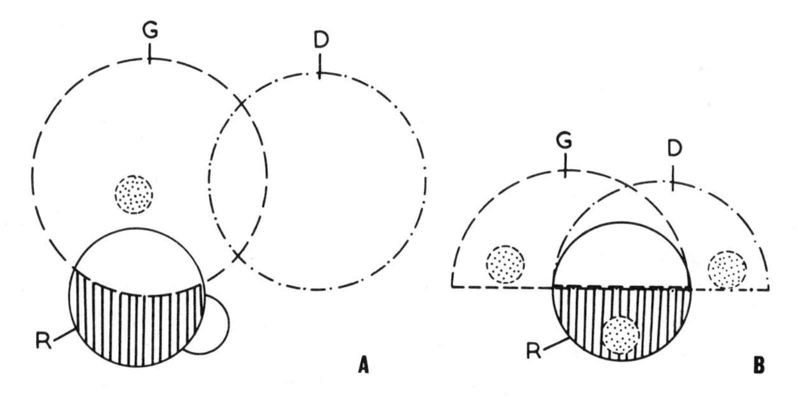

Figure 6(a). Schematic niche diagram illustrating the status of *R. acetosa* in hill grassland (cf. Figure 5). R = *Rumex acetosa*; G = grasses; D = dicotyledonous herbs. The fundamental niche of each of the components (R, G, and D) is drawn as a circle. The realized niche of *R. acetosa* is shaded vertically. The fundamental niche of establishing seedlings is shown as a dot filled circle. The size of circles and the *degree* of overlap are arbitrary.
(b). for *R. acetosella*. Symbols as (a). The fundamental niches of G and D are drawn as semicircles in order to accommodate part of R within both (Putwain and Harper, in preparation).

did not give rise to established seedlings unless components of the existing community had been removed. This confirms the impression from the previous experiment that there is no available niche for the seedling phase in the fully established and undisturbed community.

The reaction of *R. acetosella* to removal of grasses and dicotyledons from the habitat differed strikingly from the reaction of *R. acetosa*. The population of shoots of *Rumex acetosella* expanded rapidly after removal of both the grass and dicotyledonous components of the sward but not after the removal of either component separately. This presumably implies that the grass or dicotyledonous fraction of the flora was the preferential invader of the space left when the other was removed. This is expressed as a niche diagram in Figure 6b, where it is suggested that a part of the fundamental niche of *R. acetosella* must lie within the overlapping region of the fundamental niches of the grass and dicotyledonous populations. Removal of the grasses, the dicotyledonous species or the existing population of *R. acetosella,* all provided sites in which seedling establishment of *R. acetosella* could occur; seedling establishment did not occur in the undisturbed vegetation.

An essentially similar type of experiment was made on grassland by Sagar (1953) who removed grasses from communities containing abundant *Plantago lanceolata*. He was able to measure both the increase in population size by vegetative and seed reproduction, and the increase in individual size and reproductive capacity. The experiments indicated an overlapping niche between *P. lanceolata* and the grass components of the community, but did not attempt to break the community into further components.

This sort of experiment needs great care in its interpretation, and it would be extremely difficult to explain the mutual behavior of the components of such a community without a detailed knowledge of their respective physiologies and of the changes in the environment that are produced by selective elimination. The act of elimination from a mass of complex vegetation itself produces many indirect consequences. A component removed from a habitat is not only one of many species making demands on limiting light, water, and nutrient supplies; it is also a possible source of micro-environmental conditions necessary for associated species. It may, if a distateful element of the flora, protect immediately adjacent plants from the action of grazing animals (see Harper and Sagar, 1953). When killed by herbicide treatment, it remains in the community as a dried or decaying structure which in decomposition may be expected to produce changes in the nitrogen balance of the soil system, or when dry and withered will still intercept some light that could have fallen on neighbors. The herbicide treatments themselves are not perfectly selective, and care has to be taken that residual toxicity does not remain in the soil. However, bearing in mind these limitations, this type of experiment does go some way towards disentangling interrelationships in mixed vegetation.

The experiments with *Plantago* spp., like those with *Rumex* supp., indicate the powerful role of the grass flora in limiting the realized niche of the dicotyledonous species. There is abundant evidence in the agronomic literature (e.g., Tribe, 1950) that the grazing animal is a selective predator in grassland: a key to the understanding of niche diversification in grasslands almost certainly lies in the activity of the grazing animal. There is much need for population botanists who are prepared to make experiments—one of the most rewarding fields of experimentation lies at the interface of the first and second trophic levels.

LITERATURE CITED

Ashby, E., 1936, "Compound Interest and Plant Growth," *Sch. Sci. Rev.*, 67:411-423.

Black, J. N., 1960, "The Significance of Petiole Length, Leaf Area, and Light Interception in Competition Between Strains of Subterranean Clover (*Trifolium subterraneum* L.) Grown in Swards," *Austr. J. Agric. Res.*, 11:277–291.

Braun-Blanquet, J., 1951, *Pflanzensoziologie*, 2d ed., Springer, Vienna.

Brougham, R. W., 1960, "The Relationship Between the Critical Leaf Area, Total Chlorophyll Content, and Maximum Growth-rate of Some Pasture and Crop Plants," *Ann. Bot. N.S.*, 24:463–474.

Cavers, P. B., and J. L. Harper, 1967, "Studies in the Dynamics of Plant Populations. I. The Fate of Seed and Transplants Introduced into Various Habitats," *J. Ecol.*, 55:59–71.

Champness, S. S., and K. Morris, 1948, "The Population of Buried Viable Seeds in Relation to Contrasting Pasture and Soil Types," *J. Ecol.*, 36:149–173.

Charles, Allen H., 1961, "Differential Survival of Cultivars of *Lolium, Dactylis* and *Phleum*," *J. Brit. Grassl. Soc.*, 16:69-75.

Chippindale, H. G., 1948, "Resistance to Inanition in Grass Seedlings," *Nature*, London 161:65.

Clatworthy, J. N., 1960, "Studies on the Nature of Competition Between Closely Related Species," D. Phil. thesis, Univ. of Oxford.

Curtis, J. T., and R. P. McIntosh, 1951, "An Upland Forest Continuum on the Prairies-Forest Border Region of Wisconsin," *Ecology*, 32:476–496.

De Wit, C. T., 1965, "Photosynthesis of Leaf Canopies," *Versl. Landbouwk. Onderz.*, 663: 1–57.

Donald, C. M., 1963, "Competition Among Crop and Pasture Plants," *Adv. Agron.*, 15:1-118.

Ehrlich, P. R., and P. H. Raven, 1965, "Butterflies and Plants: A Study in Coevolution," *Evolution*, 18:586–608.

Greig-Smith, P., M. P. Austin, and T. C. Whitmore, 1967, "The Application of Quantitative Methods to Vegetation Survey. A. Association-analysis and Principal Component Ordination of Rain Forest," *J. Ecol.*, 55:483–503.

Harberd, D. J., 1961, "Observations on Population Structure and Longevity of *Festuca rubra* L.," *New Phytol.*, 60:184–206.

______, 1962, "Some Observations on Natural Clones in *Festuca ovina,*" *New Phytol*; 61:85-100.

______, 1963, "Observations on Natural Clones of *Trifolium repens* L.," *New Phytol.*, 62:198–204.

Harper, J. L., 1956, "The Evolution of Weeds in Relation to Resistance to Herbicides," Proc. 3d Brit. Weed Control Conf., 1:179–188.

______, 1964a, "The Nature and Consequences of Interference Amongst Plants," p. 465–462. In *Genetics Today*, (Proc. XI Int. Cong. Genet., 1963.)

______, 1964b, "The Individual in the Population," *J. Ecol.*, 52 (Suppl.): 149–158.

______, 1967, "A Darwinian approach to plant ecology. *J. Ecol.*, 55:247–270.

Harper, J. L., and J. N. Clatworthy, 1963, "The Comparative Biology of Closely Related Species. VI. Analysis of the Growth of *Trifolium repens* and *T. fragiferum* in Pure and Mixed Populations," *J. Exp. Bot.*, 14:172–190.

Harper, J. L., and I. H. McNaughton, 1962, "The Comparative Biology of Closely Related Species Living in the Same Area. VII. Interference Between Individuals in Pure and Mixed Populations of *Papaver* Species," *New Phytol*, 61:175–188.

Harper, J. L., and G. R. Sagar, 1953, "Some Aspects of the Ecology of Buttercups in Permanent Grassland," Proc. 1st Brit. Weed Control Conf., p. 256–264.

Harper, J. L., J. T. Williams, and G. R. Sagar, 1965, "The Behaviour of Seeds in Soil. I. The Heterogeneity of Soil Surfaces and its Role in Determining the Establishment of Plants from Seed. *J. Ecol.*, 53:273–286.

Huffaker, C. B., and P. S. Messenger, 1964, "The Concept and Significance of Natural Control," p. 74–117. In P. DeBach, ed., *Biological Control of Insect Pests and Weeds*, Reinhold, New York.

Macfadyen, A., 1964, "Energy Flow in Ecosystems and Its exploitation by Grazing," In D. J. Crisp, ed., *Grazing in terrestrial and marine environments*, Symp. Brit. Ecol. Soc., 4:3–20.

Obeid, M., 1965, "Experimental Models in the Study of Interference in Plant Populations," Ph.D. thesis, Univ. of Wales.

Putwain, P. D., and J. L. Harper, 1968, "Studies in the Dynamics of Plant Populations. II. Components and Regulation of a Natural Population of *Rumex acetosella* L.," *J. Ecol.* 1968.

Roberts, H. A., 1963, "Studies on the Weeds of Vegetable Crops. II. Effect of Six Years of Cropping on the Weed Seeds in Soil," *J. Ecol.*, 50:803–813.

Sagar, G. R., and C. Marshall, 1966, "The Grass Plant as an Integrated Unit—Some Studies of Assimilate Distribution in *Lolium multiflorum* Lam." Proc. 9th Int. Grassl. Cong., p. 493–497.

Takeda, T., 1961, "Studies on the Photosynthesis and Production of Dry Matter in the Community of Rice Plants," *Jap. J. Bot.*, 17:403–437.

Tansley, A. G., 1949, *The British Isles and Their Vegetation*, Cambridge Univ. Press, Cambridge.

Tribe, D. E., 1950, "The Behavior of the Grazing Animal: A Critical Review of Present Knowledge," *J. Brit. Grassl. Soc.*, 5:209–224.

Watson, D. J., 1952, "The Physiological Basis of Variation in Yield," *Adv. Agron* , 4:101–154.

Williams, W. T., and J. M. Lambert, 1959, "Multivariate Methods in Plant Ecology. I. Association Analysis in Plant Communities," *J. Ecol.*, 47:83–101.

Yoda, K., T. Kira, H. Ogawa, and K. Hozumi, 1963, "Intraspecific Competition Among Higher Plants. XI. Self-thinning in Overcrowded Pure Stands Under Cultivated and Natural Conditions," *J. Biol.*, Osaka City Univ., 14:107–129.

-11-

The Theory of the Niche

ROBERT MAC ARTHUR
Princeton University
Princeton, New Jersey

INTRODUCTION

A few words about the role of theory in ecology in general will be in order before we turn to the theory of the niche in particular. Ecological patterns, about which we construct theories, are only interesting if they are repeated. They may be repeated in space or in time, and they may be repeated from species to species. A pattern which has all of these kinds of repetition is of special interest because of its generality, and yet these very general events are only seen by ecologists with rather blurred vision. The very sharp-sighted always find discrepancies and are able to say that there is no generality, only a spectrum of special cases. This diversity in outlook, has proved useful in every science, but it is nowhere more marked than in ecology. Here, aside from a few physical, chemical, and ecological constraints, the patterns are shaped only by natural selection. The constraints set limits beyond which the system cannot wander; within these limits there is an evolutionary optimum toward which organisms should converge. But the uncertainties of the environment make the goal change from place to place and time to time, and the organisms under the influence of this selection only show a rough tendency to conform to one another and to the optimum. Clearly, no pattern of such organisms can be expected to coincide with numercial precision from situation to situation. Fortunately, numerical precision is not the only aim of science; the new hypotheses and simplification of education which come from qualitative theories can be just as rewarding.

A theory must eventually be falsifiable to be useful to a scientist, but it does not *in itself* have to be directly and easily verified by measurement. More often it is the consequences of the theory that are verified or proved false. In what follows we shall see some of the ecological constraints and some of the evolutionary optima in a disprovable form.

STRENGTHS AND WEAKNESSES OF EARLY DEFINITIONS OF NICHE

The term niche was almost simultaneously defined by Elton and Grinnell to mean two different things. To Elton, the niche was, somewhat vaguely, the animal's "role" in the community. This included its position in the food web as well as miscellaneous other habits. To Grinnell, the niche was a subdivision of the environment, a somewhat finer subdivision than the life zone. Some features of each were incorporated into the niche as Hutchinson (1958) has redefined it. In Hutchinson's niche, each measureable feature of the environment was given one coordinate in a infinite-dimensional space. The region in this space in which fitness of an individual was positive was called that individual's niche. This provided, for the first time, a definition precise enough that the term "niche" could enter into falsifiable statements. Now the statement "Two species cannot co-exist if their niches are identical" became true but trivial (since no two individuals, let alone two species, have identical niches). The statement "Two species can coexist if their niches do not overlap" becomes plausibly false, since by feeding in different places, two species would occupy non-intersecting niches even if they both depended on and competed for the same highly mobile food supply.

Hutchinson proposed to relate niche to competition by locating homogeneous patches of environment lying within the intersection of two species' niches. There, he conjectured, only one of the species would persist within an enclosure preventing migration. Although this conjecture suggests an interesting experiment which has, I believe, never been done, the experiment could not, for three reasons, falsify the hypothesis. To falsify the hypothesis we would have to know that the enclosure did indeed be within the intersection of the niches. But, as Hutchinson was among the first to realize:

1. The niche space is infinite dimensional, and we do not know how the addition of dimensions might shrink the niche intersection.
2. The niche coordinates are ambiguous, and we do not know how their alteration might change the niche intersection.
3. Where a variety of mobile resources (large and small, for example) are found in the same area, it is not clear whether that area lies in the niche intersection of a species which is efficient on large foods but able to eat small, and a second species which is efficient on small foods but able to eat large. If the area is in the intersection, then the conjecture appears false; if the area is not in the intersection, the conjecture appears untestable.

People who insist that all such terms be operational will reject "niche" just as they must reject "phenotype" and "genotype," as involving an infinite number of measurements; but some statements about *differences* between niches are perfectly testable, which is all that matters. (In fact, the parallel between "niche" and "phenotype" is more than formal. "Phenotype" embodies all the measurements which can be made on an individual during its lifetime, including those measurements which constitute its niche. Thus "phenotype" includes "niche." And to the extent that all relevant phenotype parameters affect fitness, "niche" almost includes "phenotype.") Hence, the term "niche" will mainly appear in comparative statements. We can compare niches of two similar species, or we can compare the niche of one species at two places or two times. Usually these differences will involve only one or two measurements. Levins has discussed the evolution of the niche in a fashion which I cannot improve, so I turn to other aspects.

COMPETITION AND THE STRUCTURE OF THE ENVIRONMENT

The environment, especially the terrestrial environment, has such a complex geometrical structure that it is no wonder a term like niche seems to be either ambiguous or inadequate. If animals were unable to use the structure, it would not affect the coexistence of species; but when the structure is large compared to the organisms, some analysis of the structure itself is necessary.

To help understand the structure of the environment, Levins and I introduced the concept of grain. We now call a patch of environment "fine-grained," relative to a species, if that species comes upon the resources and other components of that patch in the proportion in which they occur. Conversely, if the species can spend a disproportionate amount of its time on one resource or other component, then we called the patch coarse-grained. Thus, a warbler feeding among the tree tops in a forest possibly treats the different deciduous tree species as a fine-grained mixture and, if there is twice as much maple canopy as beech canopy, spends twice as much time in maple as in beech. Even more plausibly, the warbler comes upon the different defoliating insects in a small piece of canopy in the proportion in which they occur, so that within that piece the food supply is fine-grained. Almost certainly it sees everything within some small area reflecting the visual field of the bird. To the insects, however, especially the monophagous ones, the same canopy is obviously coarse-grained since each species may spend all of its time in one species of tree, and even in one part of the tree. A filter-feeding pelagic copepod may treat depth as a coarse-grained feature of its environment, since it selects one depth at which to concentrate its feeding, but within that depth it comes upon all particles in the proportion in which they occur. Although it only selects some of these to eat, its potential food supply is fine-grained. Finally, notice that a species may treat the environment as fine-grained even though each individual treats it as coarse. For instance, a plant species may distribute its seeds randomly over various patches of its environment, but each germinated seedling will spend its life in one patch. The sharp-sighted will, of course, object that no environment is ever perfectly fine-grained, but science is made up of such fictions. (Think how physics would be without its frictionless pulleys, conservative fields, ideal gases, and the like), and to the extent that we can approach a truly fine-grained situation the term can be useful. Of course, to make comparative statements we must have some measure of the degree of departure from fine-grainedness. (Here is one such measure: let $p_1, p_2, p_3 \ldots p_n$ be the proportions of components as they occur in nature and let $t_1, t_2, t_3 \ldots t_n$ be the corresponding numbers of independent observations of a given duration [say 5 seconds] of the organisms' time spent in these components. The sum of the t_i, T, must be fixed. Then $\sum_{i=1}^{n} \frac{(t_i - Tp_i)^2}{Tp_i}$ has a χ^2 distribution with mean $n - 1$ and standard deviation $\sqrt{2(n - 1)}$ so that

$$D = \sum_{i=1}^{n} \frac{\frac{(t_i - Tp_i)^2}{Tp_i} - (n - 1)}{\sqrt{2(n - 1)}}$$

is a measure of departure which is nearly independent of n, and its comparison is independent of the fixed number of seconds of observation.)

I now return to the problems of understanding how species coexist. To accomplish this I make the following conjectures:

(a) In a patch which is either a fine-grained mixture or homogeneous and structureless, species only coexist by virtue of resource subdivision. (Downy and hairy woodpeckers may come upon the same food items, but the downy select the small and the hairy select the large, so that coexistence is possible.)

(b) A real environment can be assembled from bricks, or building blocks, of fine-grained patches.

These conjectures, which are the basis of what I will say, require a little discussion. That a fine-grained medium cannot support more species than it has resources seems to be the message of the bottle experiments of the last thirty years. It is also rendered likely by a theoretical argument (Mac Arthur and Levins, 1964). But the usual exceptions must be made; namely, if anything keeps the populations too low to exhaust the resources, that thing may not be acting in a fine-grained fashion and could allow extras to persist. Also, of course, the definition of resource is fuzzy. To be two resources, two items of food must be independently harvestable. For instance, bark and leaves and seeds of a tree may well be three (at least) resources; but if the bark is eaten to such a level that it interferes with leaf production, it is no longer a separate resource. No amount of harvesting of seeds will interfere with leaf production, at least in the short run, and so seeds are clearly a different resource. If the first conjecture proves valid, then we only need to look at resource coordinates of niches in fine-grained environments, which is a vast improvement.

The second conjecture is more subtle. It may turn out that some, but not all, environments can be assembled from fine-grained components. If so, we will certainly need to learn how to tell which environments can. For these environments it will be quite easy to understand coexistence. I shall do this in steps, summarized in Table I.

TABLE I

Number of resources	Environment	Maximum number of species
1	fine-grained	1
2	" "	2
n	" "	n
1	coarse-grained; m-grain types	m
1 or many	coarse-grained; continuous	many
1 or many	{(coarse-grained for some species) (fine grained for others)}	?

The main new feature added by coarsening the grain is the possibility of habitat subdivision; barring other factors limiting the number of species, there can be as many species as there are grain types, even with but one resource in the system. There can also be fewer species than grain types, but sometimes this will be accomplished by one species occupying more than one grain type indiscriminately—in other words, it will treat these in a fine-grained fashion. It might feed in one or both grain types but still in a coarse-grained fashion. Here the time wasted hurrying through the poor patch types might more than compensate for feeding specializations within good patches, and a "patch generalist" would be in a position to out-compete two patch specialists. The details of when this would happen depend upon whether the specialists can travel a path which avoids poor patches, or whether the patches are more isolated. Thus some measure of patch connectedness will be called for. I have no useful measure to propose.

To make these statements testable, we return to the measure of departure from fine-grainedness. If we choose some standard departure (Say $D = K$), then there is a host of possible, and testable, hypotheses:

(a) If, for one species, $D < K$, then no other species will be present (unless they differ in resources).

(b) If, for one species, $D > K$, then other species will be present and, adding together the utilizations of all the species, $D < K$. These are just guesses, but they illustrate the machinery for testing grain statements. We might also predict that K would be less in stable environments.

Finally I have talked about coexistence within a fine-grained patch by virtue of resource subdivision and about coexistence by virtue of coarse grained utilization of the habitat. Both of these concepts can be made more numerical, with the accompanying chances of being wrong, as I now will try to show. I shall base my discussion on joint papers with Levins (Mac Arthur and Levins, 1964, 1967).

First I deal with coexistence by virtue of resource subdivision within a fine-grained patch. Resources are consumed and renewed, and the level of abundance which they reach depends upon the balance of consumption and renewal. A heavier consumption will normally lower the abundance of the resource. There will normally be some threshold level of the resource below which the consumer cannot feed rapidly enough to maintain its population. Thus, a bird may have to rear a certain number of young to compensate for its annual mortality and must gather a certain number of insects per hour in order to feed those young. This determines the threshold density of the insect resource. When the resources exceed this threshold, each consumer can gather more than enough food and its population can rise, which will cause the resource level to drop. Hence, the equilibrium level of the resource will normally coincide with the threshold level marking how few individuals the harvesters can use to perpetuate their population. (This all assumes a resource-limited consumer, of course.) This threshold level to which the consumer reduces its resources I call T. Hence, as a not-too-rough approximation, the population S_1 of a resource-limited consumer might grow according to the formula

$$\frac{ds_1}{dt} = rs_1 f(g(R_1, R_2 \ldots) - T)$$

where f is a monotonic function such that $f(o) = o$, and, where R_1, $R_2 \ldots$ etc. are the densities of the various resources in the fine-grained mixture. For pictorial purposes we consider a two-resource system, and we plot $\frac{ds_1}{dt} = o$ in the graph whose coordinates are R_1 and R_2 (see the right-hand

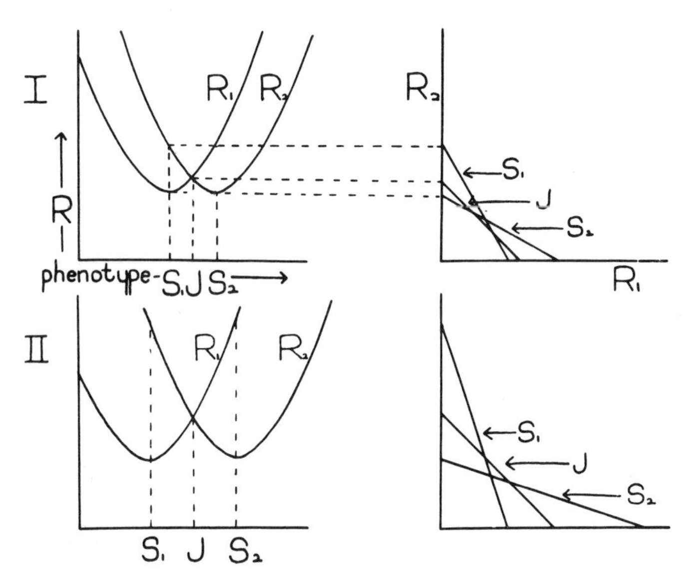

Figure 1. In the right-hand graphs, the levels to which species with phenotypes S_1, S_2, and J can reduce resources R_1 and R_2 are indicated by lines (isoclines). These lines would be curved if the resources are not alternatives to one another. In the top right-hand graph, J can invade a community containing S_1 and S_2, while in the lower right-hand graph the J line lies outside the intersection of the S_1 and S_2 lines showing J cannot invade. In the left-hand graphs, the intercepts of the isoclines of the right graphs are plotted. Here the R_1 and R_2 intercepts can be plotted as a continuous, but rather arbitrary, function of phenotype, the low point on the R_1 curve being the phenotype, S_1, which is specialized for this resource. The phenotype J is a jack-of-both-trades in the sense that it can reduce both resources equally. These graphs can be used to predict evolutionary convergence and divergence: when the resources are as different as in graphs II, J is inferior and phenotypes near J will diverge toward S_1 and S_2; when the resources are as similar as in graphs I, phenotypes S_1 and S_2 will converge toward J.

parts of Figure 1). From what we have said, $\frac{ds_1}{dt} = o$ if $g(R_1, R_2) = T$ which is the equation to plot on the graph.

In case $g(R_1, R_2) = A_1R_1 + A_2R_2$, then the resources are interchangeable and $\frac{ds_1}{dt} = o$ is a straight line as shown in the figures. If the resources provide different dietary requirements, the curve $g(R_1R_2) = T$ bends inward toward the origin and may even be asymptotic to the coordinates. At any rate, the population of resource-limited species s_1 will increase until the resources are reduced to some value along the curve $g(R_1R_2) = T$ at which time $\frac{ds_1}{dt} = o$ and there will be no further increase. We now ask whether a second and third species, with populations s_2 and J can invade. If the curve $\frac{ds_2}{dt} = o$ does not intersect $\frac{ds_1}{dt} = o$, they cannot coexist in equilibrium; but if they do intersect, equilibrium may (but need not always) be possible. If they do coexist, it must be at the resource level corresponding to the intersection of the curves $\frac{ds_1}{dt} = o$ and $\frac{ds_2}{dt} = o$. In this case, J can invade if, and only if, the line $\frac{dJ}{dt} = o$ lies inside the intersection point; and again we must decide whether J will replace both or coexist with one. (It cannot coexist stably with both because there is no value of R_1 and R_2 lying on all three lines; even if the three lines intersected at one point, it can be shown that at least one would go extinct.) Whether one or two consumer species persist depends upon whether the resource species are able to replace themselves at the level indicated by the intersection of the curves. To answer this we must also examine the equations governing the resource population growth. Both consumers will persist if $\frac{dR_1}{dt} = o$ and $\frac{dR_2}{dt} = o$ have a positive solution when the R_1 and R_2 values at the intersection are put in. The main point is that no more species than resources will persist and perhaps not as many consumers as resources. Since selection of diet is not influenced by competitions, the species, to consume different diets, must have quite different phenotypes; so the species which do coexist within a fine-grained patch should differ in phenotype. These same graphs can be related to phenotype and used to predict character divergence and convergence (see the figure), but I shall instead turn to the situation in a coarse-grained environment.

In addition to resource subdivision, the coarse-grained environment offers opportunities for habitat separation, and it is this new aspect which I shall describe here, using the same kind of graph (Fig. 2a, 2b). Now R_1 will not be one kind of resource, but rather the quantity of the mixed resources in grains of type 1. As before our graphs show us that, by virtue of the grains of habitat, no more species than grain types can persist (and perhaps fewer). And, as before, if the resources in the grains of habitat are interchangeable, then the lines will be straight. In this case, knowledge of the intercepts is sufficient, and we inquire to what level a predator can reduce its resources in each kind of habitat grain. This will be closely related to the proportion of its time the predator spends in that kind of grain.

In the figures the upper limits to the amounts of R_1 and R_2 are marked by the stippled zone. That is, the resources are only capable of renewing within the stippled zone so that only these are feasible resource levels. In

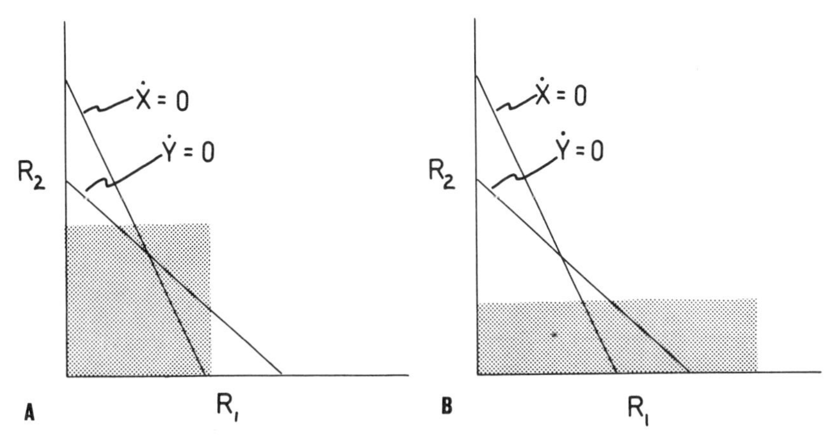

Figure 2. The levels at which the rescurces can renew themselves are superimposed with stippling onto graphs like the right-hand graphs of Figure 1. In 2a, both resources can reach the same level and the stipplied zone includes the intersection of the isoclines for species *X* and *Y*. Hence *X* and *Y* will persist. In 2b, resource 2 reaches only a very low level, perhaps because it is confined to a small part of the environment. Hence, both *X* and *Y* cannot persist and *X* alone can always oust *Y* and prevent its reinvasion.

Figure 2a, two species can coexist (unless some other limitation on resources or consumers is acting), while in Figure 2b, the inequality of resource limits prevents more than one species from persisting.

Hence, by decomposing an environment into fine-grained building blocks, it seems possible to consider separately the resource coordinates of the Hutchinson niche and the other coordinates which determine the patch classification.

THE NICHE AND SPECIES DIVERSITY

My brother and I (Mac Arthur and Mac Arthur, 1961, and later) discovered that the number of bird species breeding in a small area of rather uniform aspect could be predicted in terms of the layers of vegetation and seemed independent of the number of plant species (see Figure 3). More specifically, we let p_1 be the proportion of the total vegetation lying in the herbaceous layer between the ground and about two feet, p_2 be the proportion of vegetation in the brush layer from about 2 to 25 feet, and p_3 be the proportion in the canopy about 25 feet. Then the logarithm of the number of species was proportional to $-\sum_{i=1}^{3} p_i \ln p$ which we called the foli-

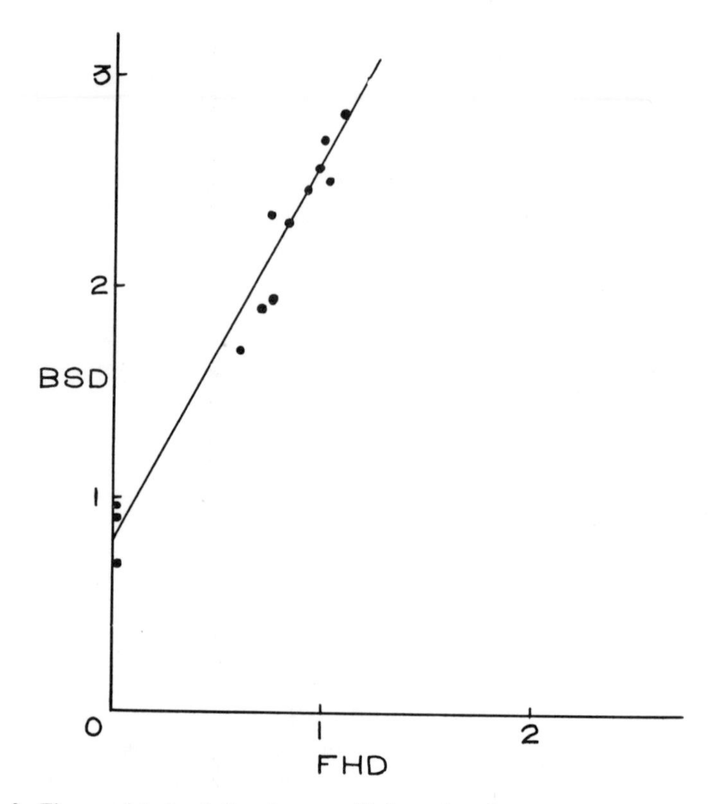

Figure 3. The empirical relation between bird species diversity, B.S.D., (roughly the logarithm of the number of species) and the foliage height diversity, F.H.D., calculated as described in the text. The bird censuses are from deciduous and grassland habitats in temperate North America.

age height diversity. In other words, the number of bird species could be predicted in terms of the structure of the habitat by means of a particular formula. (For comparisons involving different geographical areas, more than structural differences need to be considered [Mac Arthur, Recher, and Cody, 1966], and the climatic stability and environmental productivity and, perhaps, history are influential.) Here I shall show how the theory presented in the last section not only predicts the qualitative aspects of this relation but also the exact formula for foliage height diversity (although this last may be a coincidence).

We saw in Figure 2 that the number of species persisting depended upon whether the intersection of the isoclines lay in the stippled zone. And the dimensions of the stippled zone are determined by the proportion of the total area which belongs to the various patches. If patch 1 has ¾ of the area and patch 2 has ¼, then the resources in patches of type 1 should reach about three times the abundance of those in patches of type 2. In other words, the dimensions should be roughly proportional to ¾ and ¼. So the volume of the stippled area is ¾ × ¼ or, in general, a product of p_i's. Still more generally, there will be different resources in each patch type. If the patch type is absent ($p_i = o$), then the number of resources it contains must also be zero, and it seems reasonable to suppose that the number of resources is proportional to p_i. This is equivalent to a linear species-area curve, which is a first approximation to the truth. Hence the product of p_i's will contain p_1 about np_1 times, p_2 about np_2 times, and p_3 about np_3 times; that is, we will plot a graph with n times p_1 coordinates for resources in patches of type 1 and with $n \times p_2$ coordinates for resources in patches of type 2, etc. The product of all of these numbers is $p_1^{np_1} \; p_2^{np_2} \; p_3^{np_3} \ldots$ which should be related to the number of species so that the logarithm of the number of species should be determined by the logarithm of $p_1^{np_1} \; p_2^{np_2} \ldots$ which is $n \Sigma p_i \, ln \, p_i$, the formula which we have seen is actually useful. This explanation may be spurious, but for the present it serves to connect the theory of competition with the facts of species diversity. Of course, different organisms would not recognize layers of vegetation as the coarse grains of their environments, so the determination of which patches are relevant must be made empirically.

When other factors than competition limit species diversity, such as predation or history, then there is no reason to expect this theory to be relevant.

NICHE EXPANSION AND COMPETITIVE RELEASE

One of the most frequently documented bits of evidence for competition is the habitat expansion of species on islands where they are freed from their mainland competitors. For instance, Cameron (1958) pointed out that the arctic hare *(Lepus arcticus)* on Newfoundland formerly had expanded its habitat to include not only the tundra where it lives on the mainland but also the forest land throughout the island. The obvious cause was the release from competition with the snowshoe hare (*Lepus americanus*) which occupies the forests on the mainland but had not colonized Newfoundland. This example is specially interesting because the snowshoe hare was subsequently introduced and the arctic hare contracted back into its tundra stronghold. This case (like most of the others) was not as carefully studied as we might wish, but Crowell is now performing the same sorts of experiments with mice on islands off the Maine coast, and the process should soon be well-documented.

Pianka and I (Mac Arthur and Pianka, 1966), and independently Emlen (1966), gave a theoretical explanation for this kind of expansion which involves some new and testable consequences. We were concerned with optimal diet and optimal patch utilization based on economic grounds. That part of the argument which deals with niche expansion under release from competition runs as follows. The decision which patch to feed in is made before the food is located, while the decision whether to pursue and eat an item is made after the food item is located. Hence, the patch decision must be made on the grounds of expectations of yield while the diet decisions are made on the merits of the item already located. An item already located is worth pursuing if, in the time it takes to catch and eat it, no more rewarding item would be found. This decision is not based on the abundance of the item under consideration (although the rarity of other kinds of items will enter the decision), and so a reduction in the abundance of that item should not affect its acceptability in the diet. A reduction in the abundance of all items within a patch will greatly affect the expected harvest within that patch and should, therefore, change the decision of where to feed. That is, competitive reduction in food should alter the patches in which the species searches (often reducing the habitat), but should not very greatly alter the acceptable range of foods within a patch. (The diet itself, reflecting both acceptability and abundance of the foods may change.) Hence, we often expect a habitat contraction in the presence of competitors, but the range of acceptable items should be relatively unchanged. The habitat expansion is, as we have seen, frequently documented; the range of items in the diet is just a testable conjecture at this time.

The same theory leads to other testable predictions, and I include one here because optimal diet is so relevant to the study of the niche. Once the species finds an item of food, the decision of whether or not to try for it depends upon the likelihood of coming upon a more rewarding item during the time it would take to capture and eat the first one. If capture and eating take very little time compared to search (the species is a "searcher"), then the likelihood is negligible and all palatable items should be eaten. If capture and eating are relatively very time-consuming and location of other items is quick (the species is a pursuer), it will always pay to specialize. Finally, if the density of food is increased, it will cut search but not pursuit time so the species will become more of a pursuer and should specialize more. Recher (in press) has verified some of these conclusions in his heron studies.

I shall end by giving one final numerical example of how theory can suggest experiments. For this purpose I go to the Volterra (1926) equations for two consumers X_1 and X_2 and their resources R_1 and R_2, although more general equations could be used.

$$\left.\begin{aligned}\frac{dx_1}{dt} &= X_1[\alpha_{11}R_1 + \alpha_{12}R_2 - T_1] \\ \frac{dx_2}{dt} &= X_2[\alpha_{21}R_1 + \alpha_{22}R_2 - T_2]\end{aligned}\right\} \quad \dot{\mathbf{X}} = \mathbf{X}[A\mathbf{R} - \mathbf{T}]$$

$$\left.\begin{aligned}\frac{dR_1}{dt} &= \frac{r_1R_1}{S_1}[S_1 - \gamma_{11}X_1 - \gamma_{12}X_2 - R_1] \\ \frac{dR_2}{dt} &= \frac{r_2R_2}{S_2}[S_2 - \gamma_{12}X_1 - \gamma_{22}X_2 - R_2]\end{aligned}\right\} \quad \dot{\mathbf{R}} = \frac{\mathbf{r}}{\mathbf{S}}\mathbf{R}[\mathbf{S} - \mathbf{R} - G\mathbf{X}]$$

(The right hand equations are the matrix equivalents with boldface letters being column vectors and A the matrix of the alphas, and G of the gammas.)

A couple of comments are in order here. First, α_{ij} is the probability that a given item of resource j is eaten by a given individual of species i in a unit of time, and the alphas should be multiplied by w_j, the weight of resource j, if the weights vary. γ_{ij} is roughly $\alpha_{ji}S_i/r_i$. Second, I have added one new feature into the Volterra equations: the self-limitation of the resources, by means of the R_1 and R_2 terms on the extreme right. These put an upper limit (S_1, S_2) on the population of R_1 or R_2 in the absence of consumers corresponding to the stippled regions of Figure 2. There are doubtless also limits on the consumers in the presence of superabundant resources, but I will not be concerned with that case. These equations should be viewed as a plausible linear approximation to the true equations and are thus useful only near the equilibrium. That is, they are useful in studying the statics but not the dynamics of the situation. This misunderstanding has led the overenthusiastic to use Volterra equations far from equilibrium and the overcritical to assert that these equations are useless or even harmful to ecologists. I believe we should view them pragmatically, and if they can suggest novel and useful experiments they will have proved their worth. I shall use them to suggest experiments, but whether these experiments prove useful remains to be seen.

First, I shall relate the alphas to the intensity of competition. I have already done this in one way in Figures 1 and 2, where X_1 clearly outcompetes X_2 if the alphas of X_2 are so low that its isocline lies outside the X_1 isocline. I can also relate these four equations to the two classical competition equations by solving the second two for R in terms of X, at equilibrium and substituting into the first:

At equilibrium $\dot{\mathbf{R}} = \mathbf{0}$ so that $\mathbf{R} = \mathbf{S} - G\mathbf{X}$.

Substituting this into the equation for X we get

$$\begin{aligned}\dot{\mathbf{X}} &= \mathbf{X}[A\mathbf{S} - G\mathbf{X}\} - \mathbf{T}] = \mathbf{X}[(A\mathbf{S} - \mathbf{T}) - AG\mathbf{X}] \\ &= \mathbf{X}[\mathbf{K} - AG\mathbf{X}] \quad \text{where} \quad \mathbf{K} = A\mathbf{S} - \mathbf{T}.\end{aligned}$$

This is of the form of the competition equations where the competition coefficients are the terms in the symmetric matrix AG. Again we see that the alphas determine the intensity of competition. To use the equations in this form, we convert back to the long-hand version:

$$\dot{X}_1 = X_1[K_1 - \beta_{11}X_1 - \beta_{12}X_2]$$
$$X_2 = X_2[K_2 - \beta_{21}X_1 - \beta_{22}X_2]$$

where $\beta_{11} = (\alpha_{11}\gamma_{11} + \alpha_{12}\gamma_{21})$, $\beta_{12} = (\alpha_{11}\gamma_{12} + \alpha_{12}\gamma_{22})$, etc.
As is well known, β_{12} / β_{11} and β_{21} / β_{22} cannot be too large or stable co-existence is impossible.

The main question is, "How do we measure the alphas or betas in a real situation?" Here are two ways, motivated by the theory: Let K_1 be altered and we measure the change in the equilibrium populations. To alter K_1 (we don't need to know by how much it is altered) we can either go to a different environment or can act as predators ourselves, removing X_1 individials at the rate pX_1; this will reduce K_1 to $K_1 - p$.

At the former equilibrium

$$X_1 = \frac{K_1\beta_{22} - K_2\beta_{12}}{\beta_{11}\beta_{22} - \beta_{21}\beta_{12}}$$

$$X_2 = \frac{K_2\beta_{11} - K_1\beta_{21}}{\beta_{11}\beta_{22} - \beta_{21}\beta_{12}}$$

and at the new equilibrium, with K_1 replaced by K'_1, we simply substitute K'_1 for K_1 in these equations. We subtract the old from the new to get the change in equilibrium populations:

$$\Delta X_1 = \frac{K'_1\beta_{22} - K_1\beta_{22}}{\beta_{11}\beta_{22} - \beta_{21}\beta_{12}}$$

$$\Delta X_2 = \frac{K_1\beta_{21} - K'_1\beta_{21}}{\beta_{11}\beta_{22} - \beta_{21}\beta_{12}}$$

whence

$$\frac{\Delta X_1}{\Delta X_2} = - \frac{\beta_{22}}{\beta_{21}}$$

This is just the quantity we need to predict the stability of the species co-existence. (By altering R_1 we could have measured the ratio of alphas if we wished that instead.) Hence the theory tells us that by acting as pre-dators on one species and watching the change in equilibrium populations we can measure the intensity of competition. This gives the precise form of the experiment we must perform, and if the results are well correlated with our predictions (that the ratios of the β's cannot be too large), then our use of the equations will have been justified. We should not expect the observations to conform perfectly, and to the extent that they do not, we must repair the equations. This successive use and repair of the theory is, of course, they way science works.

The derivation of the competition equations has another virtue: it gives us a second recipe for calculating the coefficients from observations of the time and energy budget of the species. Thus

$$\frac{\beta_{12}}{\beta_{11}} = \frac{\alpha_{11}\gamma_{21}+\alpha_{12}\gamma_{22}}{\alpha_{11}\gamma_{11}+\alpha_{12}\gamma_{21}} = \frac{\sum_i \alpha_{1i}\alpha_{2i}\frac{w_i S_i}{r_i}}{\sum_i \alpha_{1i}^2 \frac{w_i S_i}{r_i}}$$

for instance, and each α_{ij} is a product of the proportion of time that consumer i spends where it would find resource j, times the probability that, during a unit of hunting time in this place, the consumer will actually locate, catch, and eat a particular unit of resource. This may seem pretty unmeasurable, but there is immediate hope along at least one line of attack. Closely related species which subdivide the coarse grains of the habitat and are morphologically similar need not have large differences in search and pursuit efficiency. It would seem more likely *a priori* that their efficiencies are only barely different—just enough to make each superior in its own patch type. In this case only the time budgets enter into the calculation of $\frac{\beta_{21}}{\beta_{11}}$, and the limiting similarity of coexisting species can be related to the time budget.

In this case, if r, S, and w terms are equal

$$\frac{\beta_{12}}{\beta_{11}} = \frac{t_{11}t_{21}+t_{12}t_{22}}{t_{11}^2 + t_{22}^2}$$

or more generally

$$\frac{\beta_{ij}}{\beta_{ii}} = \frac{\sum_k t_{ik}t_{jk}}{\sum_k (t_{ik})^2}$$

For example, the number of seconds the warblers feed in different parts of a spruce tree can be used to give preliminary equations (X_1 = myrtle, X_2 = Black-throated green, X_3 = blackburnian, X_4 = bay-breasted), with each equation divided by β_{ii} .

$$\frac{dx_1}{dt} = r_1 X_1 [6{\cdot}190 - X_1 - {\cdot}490\, X_2 - {\cdot}480\, X_3 - {\cdot}420\, X_4]$$

$$\frac{dX_2}{dt} = r_2 X_2 [9{\cdot}082 - {\cdot}519\, X_1 - X_2 - {\cdot}959\, X_3 - {\cdot}695\, X_4]$$

$$\frac{dX_3}{dt} = r_3 X_3 [6{\cdot}047 - {\cdot}344\, X_1 - {\cdot}654\, X_2 - X_3 - {\cdot}363\, X_4]$$

$$\frac{dX_4}{dt} = r_4 X_4 [9{\cdot}014 - {\cdot}545\, X_1 - {\cdot}854\, X_2 - {\cdot}654\, X_3 - X_4]$$

Here the coefficients of the X's are given by equation (1) using the data from Figure 4, and the constant terms in the brackets come from knowing the populations of the species. For instance, since there were 2 pairs of myrtles per 5 acres, 5 of black-throated green, 1 of blackburnian and 3 of bay breasted, we get $K_1 = 2 + 5 \times \cdot 490 + 1 \times \cdot 480 + 3 \times \cdot 420 = 6 \cdot 190$. From these equations we can calculate nearly anything we wish. Here I shall illustrate by finding the values of K_1, K_2, and K_3 which will prevent the bay-breasted warbler, X_4, from invading a community containing only the first three. For this purpose I abbreviate the matrix of X_1, X_2, X_3 coefficients by B:

$$B = \begin{bmatrix} 1 & {\cdot}490 & {\cdot}480 \\ {\cdot}519 & 1 & {\cdot}959 \\ {\cdot}344 & {\cdot}654 & 1 \end{bmatrix}$$

so that the equilibrium K values of the first three species are given by the column vector

$$\begin{bmatrix} K_1 \\ K_2 \\ K_3 \end{bmatrix} = B \begin{bmatrix} X_1 \\ X_2 \\ X_3 \end{bmatrix}$$

We know the bay-breasted can invade only when $K_4 - \Sigma \beta x > o$ for small X_4 and the values of X_1, X_2, and X_3 determined by equations (2). That is, it can invade when

$$K_4 > \begin{bmatrix} {\cdot}545 & {\cdot}854 & {\cdot}654 \end{bmatrix} \begin{bmatrix} X_1 \\ X_2 \\ X_3 \end{bmatrix} = \begin{bmatrix} {\cdot}545 & {\cdot}854 & {\cdot}654 \end{bmatrix} (B^{-1}) \begin{bmatrix} K_1 \\ K_2 \\ K_3 \end{bmatrix}$$

and, performing the inversion of the matrix B and premultiplying by the row vector, I get as the condition for invasion:

$$K_4 > {\cdot}1392\, K_1 + 1{\cdot}0775\, K_2 - {\cdot}4461\, K_3$$

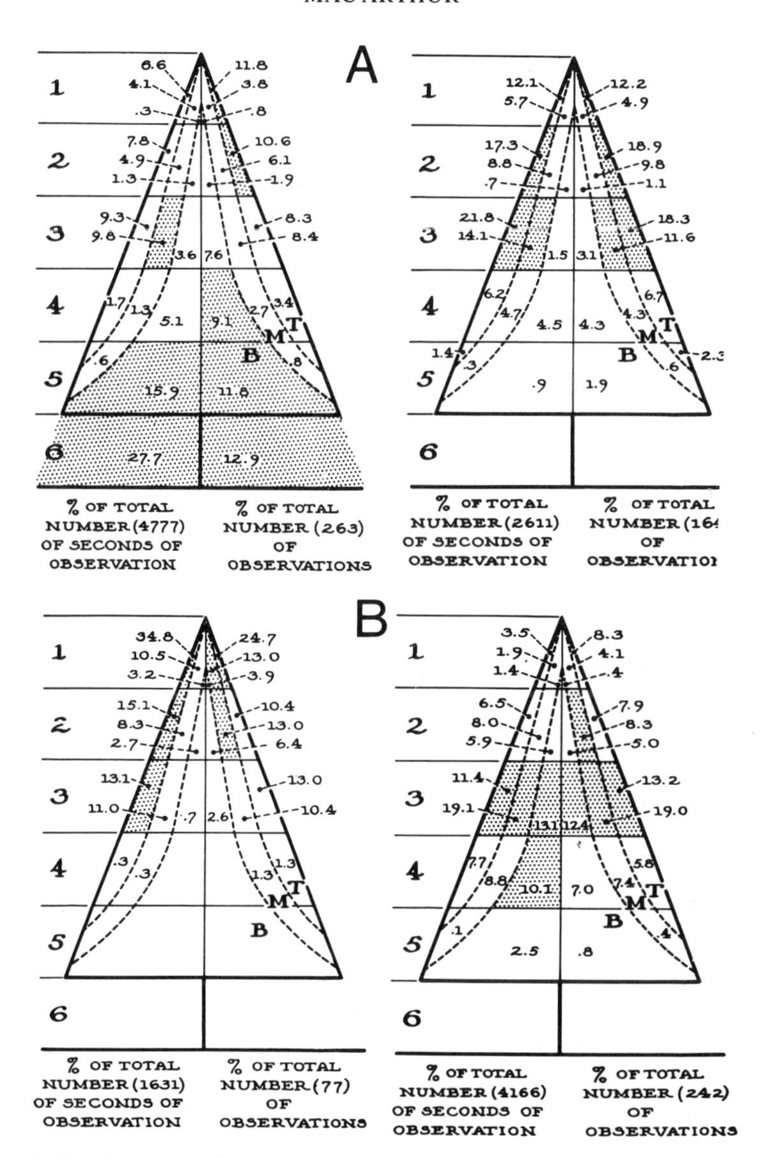

Figure 4. The figures are diagrammatic spruce trees subdivided into zones, with the time budgets of four species of warblers entered into the zones. From upper left to lower right, the species are myrtle, black-throated green, blackburnian, and bay breasted. The shaded zones indicate the most concentrated activity (from Mac Arthur, 1958).

When the distributions of foods are altered so that the K's no longer satisfy this inequality, we expect the bay-breasted warbler to be absent from the community. Notice that the negative term in the equation shows that an increase in blackburnian warblers should, by itself, make invasion easier for the bay-breasted. There are a great many testable predictions hidden in the equations. For instance, in the linear approximation on an island covered with similar forest but lacking all but the black-throated green, the density of this species should rise from 5 pairs per 5 acres to 9.082 pairs. The K's can be independently estimated from the time budget (from the formula $K = As - T$) so that $K_1 = \sum_j a_{ij} S_j - T_1$. By our assumptions, S_j may be roughly equal to the volume of the patch times the resource density within it and a_{ij} proportional to the proportion of its time the consumer spends within that kind of patch). These estimates can be directly compared with the other estimates from the censuses. A more accurate analysis of the warbler data would involve selecting feeding areas of equal S.

ACKNOWLEDGEMENTS

Most of this work is a result of studies carried out jointly with R. Levins. Drs. H. Horn and E. Leigh provided useful criticisms.

LITERATURE CITED

Cameron, W. Auston, 1958, "Mammals of the Islands in the Gulf of S. Lawrence," *Nat. Mus. Canada, Bull.*, 154.

Emlen, J. Merrit, 1966, "The Role of Time and Energy in Food Preference," *Amer. Natur.*, 100:611–617.

Hutchinson, G. E., 1958, "Concluding Remarks," Cold Spring Harbor Symp. Quant. Biol., 22:415–427.

Mac Arthur, R. H., 1958, "Population Ecology of Some Warblers of Northeastern Coniferous Forests," *Ecology*, 39:599–619.

Mac Arthur, R. H., and R. Levins, 1964, "Competition, Habitat Selection and Character Displacement in a Patchy Environment," *Proc. Nat. Acad. Sci.*, 51:1207–1210.

——., 1967, "The Limiting Similarity, Convergence, and Divergence of Coexisting Species." *Amer. Natur.*, 101:377–385.

Mac Arthur, R. H., and J. W. Mac Arthur, 1961, "On Bird Species Diversity," *Ecology* 42: 594–598.

Mac Arthur, R. H., H. Recher, and M. Cody, 1966, "On the Relation Between Habitat Selection and Species Diversity," *Amer. Natur.*, 100:319–332.

Mac Arthur, R., and E. Pianka, 1966, "On Optimal Use of a Patchy Environment," *Amer. Natur.*, 100:603–609.

Volterra, V., 1926, "Variazione e Fluttuazione del Numero d'Indvidui in Specie Animali Conviventi," *Mem. Accad. Naz. Lincei*, 2:31–113.

-12-

When are Species Necessary?

G. E. HUTCHINSON
Yale University
New Haven, Connecticut

From the standpoint of the most fundamental kind of numerical taxonomy, any specimen studied can be represented by a point in an n-dimensional space, which, following Silvestri and his colleagues (Silvestri, Turri, Hill, and Gilardi, 1962) we may call a *taxonomic space*.

The number of dimensions employed is the number of numerically estimable variables used as taxonomic characters. These should be as numerous and as quantitatively diversified as possible, following the Adansonian principle. In practice, some provision against using great numbers of highly correlated characters is probably often needed; but this makes no difference in the development of the model now presented.

Ordinarily the taxonomist is concerned with two quite distinct things. the distance between points in the taxonomic space and the tendency of the points to form clusters. These two aspects of taxonomy are quite different. In the first we are concerned with deciding which of a series of known organisms a newly discovered organism is most like. This is an entirely different question from that of the form of the probability density of points throughout a given region of the taxonomic space.

The first process is regularly employed as a practical procedure in medical bacteriology. It may be of great importance to a patient that a particular strain of bacterium isolated from his body is so like something familiarly known and named that a treatment based on previous experience with the named strain has a reasonably high probability of success. It would of course make treatment easier if every strain to be determined were so like something already known that the favorable prognosis after treatment was almost certain. This would imply clustering; it may well not exist, and yet, if enough points have been studied in the past, any new one will be close enough to a known one to make previous experience useful in therapy.

The second aspect of taxonomy, the recognition of clusters, is only appropriate to cases where clusters exist. It has become clear in recent years (Cowan, 1962, and other papers in the same excellent Symposium on Microbial Classification) that part of the difficulty of the taxonomy of many groups of organisms is due to the forcing of a system, predicated on the existence of clusters, onto groups of organisms which taxonomically are not separable into discrete groups. Given certain restrictions about age, sex, and other kinds of polymorphism, clusters, when they exist, may be reasonably equalled with the "good species" of the classical taxonomy of higher plants and animals.

In order to indicate how this sort of taxonomy may operate, I want to give an example based on the planktonic rotifers of the genus *Filinia*, on which I have recently (Hutchinson, 1964) done a little work. The mode of procedure may appear naive to those accustomed to the more sophisticated techniques of numerical taxonomy, but it happens to lead very conveniently into further theoretical developments.

Four taxa, which seem to be good species, or to be represented by clusters, are apparently involved.

1. *Filinia longiseta* (Ehrenberg), body ovoid, with moderately long appendages, the posterior one inserted well in front of the posterior end of the body. Said to be a vernal and autumnal species in ponds in temperate regions.
2. *F. limnetica* (Zacharias), ovoid, larger than *longiseta* with relatively longer appendages, the posterior one inserted well in front of the posterior end of the body. A summer form in the epilimnia of temperate lakes, probably perennial in subtropical waters.
3. *F. terminalis* (Plate), ovoid, appendages moderately long, the posterior one inserted very close to, or at, the posterior end of the body. A cold water form, in winter intemperate lakes, persisting in the hypolimnion in summer, or at great altitudes in subtropical latitudes.
4. *F. pejleri* (Hutchinson), spindle-shaped appendages moderately long, posterior one broadly inserted at the posterior end of the body. A warm water form widespread in India and South Africa and recorded from Arizona.

Numerous other characters could doubtless be added, such as the number of mastax teeth or of ovarian nuclei. Far more work on these animals would be desirable; but for the moment the metrical characters of the body, the appendages, and their insertion are sufficient to illustrate the procedure.

In Figure 1, the approximate outlines of the clusters of points representing these four species are projected onto a variety of planes. The figure is tentative and diagrammatic, though it utilizes 70 points for *terminalis* and 41 for *limnetica*. In 1(a) the distinction of *pejleri* as more spindle-shaped

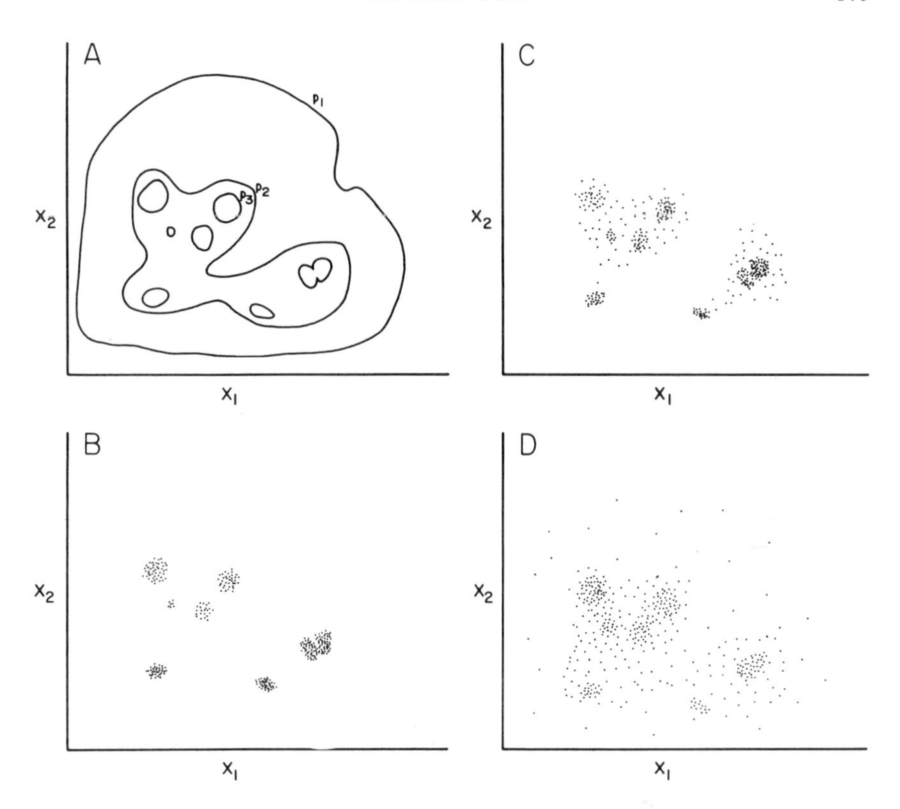

Figure 1. Projections of the *n*-dimensional envelopes enclosing points defining four species of *Filinia* onto various planes. The resulting two-dimensional envelopes are approximate only as the number of published measurements for some variables is quite inadequate.

than the others is shown; in 1(b) *pejleri* and *terminalis* are separated from the other two species by the position of the insertion of the posterior appendage; in 1(c) *longiseta* and *limnetica* appear separated by the greater length of the appendages in the latter, though occasionally there is evidence of intermediate forms which seem more likely to be due to local introgressive hybridization than the widespread existence of transitional individuals (Hutchinson, 1964). Considering all the characters, it is clear that in the five-dimensional space employed ideally in the analysis, each species is represented by a genuine discrete cluster, even though there are a few scattered points lying between *longispina* and *limnetica.*

Figure 1(d) employs an ecological datum, namely, the temperature of the water in which the species occurred (there are no reliable data for *longiseta,* but more temperature data than measurements are published for the other species). Since the specimens were living at the recorded temperatures, these temperatures must be within the tolerances of the organisms; if enough data has been collected, the range along the axis for temperature should be an estimate of the temperature tolerance of the species, and so a physiological character. In the present case, the character is probably a simple one, not necessarily operating through competitive relationships. Carlin (1943), in his classic work on the rotifers of the Motala River system, found *limnetica* to occur sporadically only in late summer, long after *terminalis* had disappeared. It is probable that if a series of envelopes for oxygen tolerance could be drawn, these would look rather like those for temperature, though with more overlap by *terminalis*. The diagram is, of course, not needed to show that we can use physiological properties whenever they can be measured as defining axes in the taxonomic space. This is naturally regularly done in microbiology. The inclusion of physiological characters, however, provides an easy transition to the next paragraph.

We now return to the general model. We consider only those axes which are genuinely independent. This restriction will not reduce the discreteness of such clusters as exist. We suppose that every independent axis represents a character that is in some sense adaptive. If it is not, we inquire whether some other character, a pleiotropic effect of the same genes as produce the adaptively neutral character, may not exist. The only other cases that are likely to give trouble are characters solely involved in maintaining the integrity of the cluster (secondary sexual characters of all sorts, so properly dear to the hearts of entomologists). Since in the critical cases to which we will apply the model this contingency does not arise, we need not worry about it in the present context. The argument from now on does rest on the faith that with this apparent exception all characters are either adaptive or more rarely represent other adaptive characters; to the present writer this seems extremely probable (Hutchinson, 1966). The various clusters, if they exist, will therefore represent adaptive peaks in the original sense of Wright (1932).

We now suppose that all adaptation, except that of the environmentally arbitrary intragroup type functioning to maintain the stability of the clusters, is in a certain sense adaptation to the properties of the environment. More specifically, we suppose that for every axis a function exists which transforms the values of the taxonomic variable measured along the axis into equivalent values of an environmental variable.

In the simplest case, larger animals can usually eat larger food than related smaller animals. We could, therefore, using suitable discretion, (cf., Ashmole and Ashmole, 1967), substitute for the appropriate dimension of the animal or its trophic structures, a function of the size of the food (maximum, mean, mode, or something more elaborate).

Lengths of organs responsible for heat loss, as the tails of certain rodents, could be similarly substituted by a function of the thermal properties of the environment. In the case of *Filinia*, appendage length may have something to do with the degree of planktonic habits, and so of water depths. Physiological characters, such as temperature tolerance, naturally lend themselves directly to the process of transformation, the result of which is to convert the taxonomic space T into the niche space N defined in an earlier contribution (Hutchinson, 1957).

It is to be noted that if the transformation is achieved by means of exclusively monotonic functions, the clusters will maintain their identities, not overlapping each other. At first sight it would seem, however, that in the new space N these boundaries would be very much closer together than in T. A very large class of points in T would, if clusters are formed and adaptive peaks exist, represent poorly adaptive genomes, while in N provided the environmental extremes are avoided, all points should represent habitable environments. It is, however, always possible that a number of points in N represent environmental conditions that do not happen to be present in the biotope under consideration. We return to an aspect of this matter later in the discussion, noting here only that such considerations suggest that the transformation functions would not be particularly simple.

Going back to the taxonomic space T it is obvious that it is to be looked on as a probability space in the sense that given a certain large number of points, these will be distributed throughout the space according to a probability structure. If clusters are formed, the probability density is high in the clusters and lower outside them; if clusters are not formed, there is a large continuum of moderate probability density. In Figure 2(a) contours are drawn within which the probability density is greater than some specified values $p_3 > p_2 > p_1$; if p_3 is large and practically no points are expected outside the p_3 contour, we have pronounced cluster formation as in 2(b); if the differences in probabilities are small, we have a continuum as in 2(d), while an intermediate assignment of probabilities gives an intermediate condition, 2(c).

Figure 2(b) obviously corresponds to the usual condition in well known taxocenes of ordinary bisexual metazoa such as birds or most insects, for which, in any given fauna, there is only a very small proportion of doubtful or difficult species.

Figure 2(c) corresponds to the situation in "difficult" apomictic genera of flowering plants. In the present example a "splitter" would recognize at least seven species, a "lumper" two; a middle-of-the-road taxonomist would probably have some reservations but would end by recognizing four, though he might decide to work on some other genus.

Figure 2(d) corresponds to a taxonomic continuum of a kind that is now

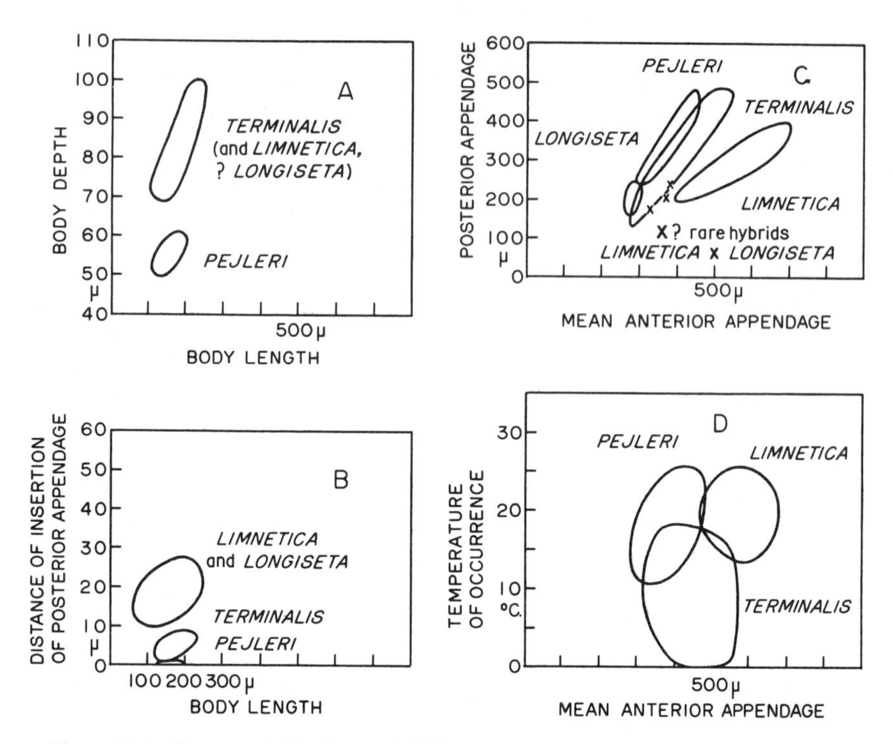

Figure 2(a). Contours indicating probability of occurrence of individuals characterized by particular values of the variables X_1 and X_2. (b). Hypothetical distribution of individuals when p_3 is much greater than p_2; seven "good" species, one showing dimorphism, are present, with only one doubtful point which could be a hybrid, a mutant or an unrecognized very rare species. (c). The same with p_3 greater than p_2 but the latter higher than in B, giving the kind of situation found in "difficult" genera; a splitter would recognize seven or eight species, a lumper two. (d). p_3 little greater than p_2 giving a continuum with vaguely defined clusters, of a kind believed to occur in some groups of procaryotes.

often believed to be characteristic of procaryote organisms. Some clustering is apparent; and since the individual points of clusters are often likely to provide samples since they are numerous, some clusters are likely to get names, as will also some points from relatively low density regions.

In terms of ordinary genetical evolutionary theory, the continuum characterizes simple organisms with little sexual exchange. Evolution primarily involves mutation; and most mutations with positive adaptive value involve little interaction, so that within the general character range of the continuum, most points represent viable genomes.

The extremely discrete speciation of 2(b), on the other hand, is ordinarily supposed to be typical of dioecious eucaryote organisms, in which each cluster represents a genetically, very well integrated adaptive peak which is inevitably impaired by hybridization. Regions between the clusters thus correspond to less adapted genomes, and the biology of the organisms is adjusted to maintain isolation.

The situation expressed by 2(c) represents a condition that in a limited number of cases in higher plants, and probably in a few animals also, is extremely successful. It has been the subject of much discussion among botanists, (e.g., Clausen and Hiesey, 1958; Heslop-Harrison, 1959), but it is difficult as yet to isolate any general principle applicable to both the plant and animal cases. All of these three situations should, as we have seen, have some counterpart in the structure of the niche space.

We may now return to the rotifers to see how far they accord with the elementary theoretical expectations just put forward. The group is divided into three classes: the Seisonaceae, the Bdelloida, and the Monogononta. The Seisonaceae contain a single valid genus with at least two species of ordinary bisexual animals; they are marine and epizoic on the crustacean *Nebalia*. Other than indicating the possibility of ordinary dioecious sexuality in the subphylum, they are of no further concern in the present context. The Bdelloida comprise four families with twenty genera and about three hundred species of animals, usually living in excessively small bodies of fresh water, such as the pellicle collecting at the bases of the leaflets of damp mosses or otherwise associated with these plants. The group contains species capable of extreme anabiosis. All species are exclusively parthenogenetic, and no males have ever been recorded. The Monogononta are divided into three orders with 18 families, nearly 100 genera, and well over 1,000 species. It is primarily found in the littoral benthos of ponds and lakes with a striking development of taxa in special habitats such as wet sand or *Sphagnum*; a number of species are secondarily marine. Many species are euplanktonic. Typically, numerous parthenogenetic generations occur until the production of a generation of mictic females. These produce males parthenogenetically, but many are then fertilized by these males, and lay resting eggs.

The taxonomy of the bdelloid rotifers depends almost entirely on the careful observation and delineation of living, fully expanded feeding specimens, no certain method of fixation that preserves the systematic character having been discovered. In spite of the absence of recognizable type specimens and comparative museum material, the taxonomy of the group seems quite well understood; to judge from the most recent literature, bdelloid species are well-defined entities. The group has long fascinated amateur microscopists, and it seems as though the experienced bdelloid watcher, provided he has access to the literature (Bartos, 1953; Voigt, 1957), should have no more difficulty in identifying what he sees through his microscope than the experienced bird watcher has with what he sees with his binoculars. There are admittedly many doubtful and inadequately described species, but this is almost certainly due to inadequate observation and not to any inherent difficulty in the taxonomy when the characters are properly observed.

The taxonomy of the Monogononta is usually somewhat easier; many, though not all, species can be adequately preserved, mounted, and compared with subsequently collected specimens. In general, it appears that the littoral benthic species are well defined, and that when different workers investigate the same genus or family, they come to essentially the same conclusions. A few difficult groups of species doubtless exist in some large genera such as *Cephalodella*, but this is often the case in a small minority of taxa in large genera up and down the animal kingdom. Among the planktonic genera there seem to be rather more of these difficult cases; *Filinia longiseta* and *F. limnetica* perhaps provide an example, and there are others in *Brachionus* and perhaps in *Polyarthra*. In the genus *Keratella*, the most abundant in individuals of all genera of freshwater metazoa in temperate regions, a situation superficially resembling that in difficult apomictic genera of flowering clearly exists.

If the existence of bisexuality is a necessary condition for the evolution of "good" species, we would have to assume that all the lines ancestral to such species in the Bdelloida were at least at times bisexual, and developed parthenogenesis separately late in their evolution. This seems fantastically unlikely. It is far more probable that part of what Mayr (1957) wrote about the apparent occurrence of some discrete species in procaryote organisms applies also here. "Curiously enough there seem to be a number of discontinuities which make taxonomic subdivision possible. The most reasonable explanation of this phenomenon is that the existing types are the survivors among a great number of produced forms, that the surviving types are clustered around a limited number of adaptive peaks, and that ecological factors have given the former continuum a taxonomic structure." We have seen how it is reasonable to pass from the taxonomic space T to the niche space N, any cluster in T corresponding, in a single sympatric assemblage to a realized niche in N. If we are to adopt the suggestion made by Mayr and supported by the case of the bdelloid rotifers, we seem forced to conclude that complex environments somehow subdivide naturally in one of a number of predetermined possible ways, as if they were intersected by cleavage planes. When we look at rotifers, it is hard to avoid the conclusion that each habitat of bdelloids or littoral benthic genera of Monogononta is likely to provide a considerable but finite class of niches, each itself an abstraction from a class of more or less similar discrete geometrically defined microhabitats. If these niches were represented by discrete regions in n-dimensional space, it is reasonable to suppose that considerable vacant regions, not representing conditions realized in nature, might exist between the niches. Supposing we consider a species of rotifer found in the axils of leaves, which make an angle between 30° and 40° with the stem, while a related species live on the free stem. The angle defining the shape of the habitat would be between 30° and 40° in one case, 180° in the other. If no projections giving angles of, say, 60° or 90° occurred in the habitat, part of the niche space would be empty. When, as in this case, the geometrical arrangement of habitats provides discontinuities, we might expect to find the organisms inhabiting them also showing discontinuities. In the plankton, the only probable discontinuities are likely to involve food sizes; and in view of the general instability of the phytoplankton, the sizes of the available foods may vary greatly from time to time and place to place. In such circumstances, a limited development of a taxonomic continuum, interrupted at intervals by amphimixis, might prove advantageous.

The freshwater Gastrotricha of the subclass Chaetonotoidea are exclusively parthenogenetic, like the Bdelloida, and again appear to be discretely speciated. In the Cladocera we have life histories superficially, though not cytologically, like those of the Monogononta; here again the really difficult genus *Bosima* is largely planktonic. It is clear, therefore, that the argument does not apply only to rotifers. It is likely to be important in diatoms, and I suspect that conversations with Dr. Ruth Patrick were the original elements of this paper.

In view of the fact that other organisms inevitably are very significant objects in the environment of any given organism, and that these other organisms are structured, and if eucaryote, usually speciated, it would seem likely that even if speciation is not required to reduce the incident of uneconomical genetic recombinations, it will nevertheless develop in response to the structured and often speciated living part of the environment.

The analysis which we have just made may seem in retrospect obvious and even trivial. It does, however, suggest that in some groups of parthenogenetic or apomictic organisms it might be possible to use the ratio of "difficult" to "good" species as a measure of the release of the pressure to produce discrete species imposed on organisms by the structure of their environments. The results of such an approach, even if it proved practically very difficult, would be of considerable interest.

LITERATURE CITED

Ashmole, N. P., and M. J. Ashmole, 1967, "Comparative Feeding Ecology of Sea Birds of a Tropical Oceanic Island." *Bull. Peabody Mus. Nat. Hist.*, 23.

Bartos, E., 1951, "The Czechoslovak Rotatonia of the Order Bdelloidea," *Vest. Csl. Zool. Spol.*, 15:241–500.

Carlin, B., 1943, Die Planktonrotatorien der Motalaström, *Meddel. Lunds Univ. Limnol. Inst.*, 5.

Clausen, J., and W. M. Hiesey, 1958, "Experimental Studies on the Nature of Species IV. Genetic Structure of Ecological Races." *Carnegie Inst. Pub., No. 615.*

Cowan, S. T., 1962, "The Microbial Species—a Macromyth?" Symp. Soc. Genet. Microbiol., 12:433–455.

Heslop-Harrison, J., 1959, "Apomixis, Environment and Adaptation." In *Recent Advances in Botany* IX. Int. Bot. Congr., Montreal, 1959, Vol. 1. pp. 891–895.

Hutchinson, G. E., 1957, Concluding remarks. Cold Spr. Harb. Symp. Quant. Biol., 22:415–427.

______. 1964, "On *Filinia terminalis* (Plate) and F. *pejleri* sp.n (Rotatoria: Family Testudinellidae)," *Postilla*, 81.

______. 1966, "The Sensory Aspects of Taxonomy, Pleiotropism and the Kinds of Manifest Evolution." *Amer. Natur.*, 100:533–539.

Mayr, E., 1957, "Difficulties and Importance of the Biological Species Concept." *Publs. Amer. Ass. Adomt. Sci.*, 50:371–388.

Silvestri, L., M. Turri, L. R. Hill, and E. Gilardi, 1962, "A Quantitative Approach to the Systematics of Actinomycetes, Based on Overall Similarity." In *Microbial Classification*, Symp. Soc. Gen. Microbiol., 12:333–360.

Voight, M., 1956–57, *Rotatoria: Die Rädertiere Mitteleuropas* I Textband, II Tafelband, 115 pl. Berlin. Gebr. Borntraeger.

Wright, S., 1932, "The Roles of Mutation, Inbreeding, Crossbreeding and Selection in Evolution." *Proc. 6th Int. Conf. Genet.*, I:356–366.

-13-

Toward a Predictive Theory of Evolution

L. B. SLOBODKIN
University of Michigan
Ann Arbor, Michigan

THE INQUIRING ANIMAL AND THE RETROSPECTIVE THEORETICIAN

I will be concerned with animals that might come to an evolutionary biologist for his advice on their state of evolutionary health. The animals may be concerned about some prospective change in their anatomy or physiology or habitat and might like to know how this influences their probability of evolutionary success. Obviously, no animals are going to behave in this way. Nevertheless, after more than a hundred years of development of evolutionary theory, this type of question ought to be answerable if asked. On a very practical level, the obverse question is being presented with frightening insistence by the environmental changes that result from human activity. Clearly, environmental changes produce biological and, in particular, evolutionary consequences; and it is highly desirable to be able to predict these consequences.

If we are free to consider suitably simplified animals, environments, and questions we can, in fact, provide tentative answers to hypothetical inquiring animals. For example, other things being equal, a higher reproductive rate is evolutionarily preferable to a lower one. It is more difficult to specify what other things must be kept equal.

In the subsequent discussion, I will attempt to avoid either simplifying the problem or restricting my discussion to situations in which other things are equal. This will limit the strength of my conclusions but will preserve their range of applicability.

There already exists a highly developed mathematical theory of population genetics, which can predict, in fair detail, the changes and constancies in gene frequencies that will result given certain selective values, breeding systems, population sizes, and the spatial and temporal distribution of these properties. To a large extent, the theory rests on the ability to measure selective advantage. This can be done by waiting several generations and then determining the relative number of individuals in the population which are descended from two individuals or genotypes or alleles. The greater the relative number of descendants, the greater the selective advantage, other things being equal. This measurement procedure is rigorous, unambiguous, and precisely what is required for the development of the formal theory. It does not, unfortunately, permit an immediate reply to the inquiring animal (cf., Medawar 1960).

An alternative, less rigorous, method of measuring selective advantage is possible. If there is reason to believe that some particular anatomical or physiological properties will be of advantage to an organism, and if the relation between these properties and particular alleles is known, then a selective value could be assigned to alleles based on the expected differential production of descendants. This involves a series of assumptions about the interaction between the biological features of the organisms and the present and future environment. The quality of the numerical estimate of selective advantage, derived in this way, will be no better than the quality of the information that morphologists, physiologists, and ecologists can provide. My analysis will not change this fact, but may help by defining certain relations that may be expected to exist between the properties of organisms and the properties of their environment.

The situation would be considerably eased if it were the case that there had been definite trends during the history of evolution. One could then consider that whatever had shown a consistent increase in evolution was at least a strong correlate of fitness, if not identical with it. This would permit an assessment of the relative selective advantage of two individuals by their relative measurements of whatever it was that showed a consistent trend. Unfortunately, there is no single property of organisms that has shown a consistent increase during the course of evolution. All properties of organisms show different trends in different phyletic lines. Even such likely-sounding candidates as health, longevity, fecundity, and complexity of organization fail to behave consistently (Slobodkin, 1964).

The problem, therefore, is how to develop a predictive theory, suitable to provide answers to an inquiring animal, in the face of an apparent absence of directly measurable, reliable, non-retrospective measures or correlates of fitness and selective advantage.

It is also very clear that, despite the general validity of mechanistic approaches to biology, we can not predict evolutionary success on biochemical criteria alone. That is, there is no single best biochemistry for organisms in general. This can be demonstrated by constructing a list of pairs of closely related species such that one member of each pair is a living species and the other member is extinct. Obviously the biochemical similarity is greatest between members of each pair; but, by any reasonable definition of evolutionary success, all the living species are evolutionarily successful despite their biochemical differences.

DOES A PREDICTIVE THEORY EXIST?

It is possible to argue that in fact there simply is no possibility of developing a causal theory of evolution with any predictive power worth considering. The admittedly stochastic events which combine to produce evolutionary change may make interesting prediction impossible. This argument cannot be dismissed out of hand, despite the fact that I find it repulsive in that its acceptance would involve cutting off an apparently legitimate empirical question from further investigation and in that sense a kind of intellectual despair.

The difficulty in producing a predictive theory of evolution is that:

1. While a rigorous definition of selective advantage exists, it is largely retrospective.
2. There is no unique historical trend in the evolutionary process to date.
3. Biochemical similarity between organisms does not, in itself, imply that they will also be similar in evolutionary success.

Given these considerations, I must proceed to demonstrate that a predictive theory is possible in principle and specify the theory in operational terms. The argument will consist of three parts. I will first present a metaphor of an "existential game," designed to provide insight into evolutionary processes. I will then make a series of empirical statements about how organisms do, in fact, respond to certain environmental events; and finally, I will present a series of empirically testable statements about how organisms will evolve and how the hypothetical inquiring animals may be given useful answers.

THE METAPHOR OF THE "EXISTENTIAL GAME"

The metaphor of a game between organisms and their environment is immediately suggested by the delicacy of specific adaptive mechanisms in organisms, and several authors have considered the possibility of describing the interaction between organisms and their environment in terms of formal game theory (Lewontin, 1961; Warburton, 1967; Slobodkin, 1964; Bonner, 1965). Various attempts have been made to define the biological analogues of utility functions, payoffs, strategies, and the other elements that enter into the definition of a game in the sense of formal game theory (Rapoport 1960). These attempts have not been as productive as might have been anticipated. I believe this lack of success has not been due to lack of ingenuity on the part of investigators but is due to a fundamental metatheoretical property of formally analyzable games. Specifically, games are played on a game board, playing field, or arena. This is not a trivial point but is absolutely central to our discussion.

Which players win and which players lose is determined on the playing field, but the value or utility of their winnings is determined away from the playing field. The winner of a poker game is paid in money, not cards, and the winner of a tennis match may get a silver cup which may have ornamental value in his house but is useless on a tennis court. Montaigne (1934) reports, as a humorous story, that a man perfected himself at throwing millet seeds through the eye of a needle and was rewarded for his dexterity by his king with several needles and a few bushels of millet seeds. Huizenga (1949) discusses at length the social implications of the existence of playing arenas. The fact that playing games may be a pleasant activity is irrelevant from the standpoint of formal game theory.

If an analogy exists between evolutionary processes and games, it must be with some game in which the players can never leave the arena. Evolutionary success involves playing a never ending game with the environment to the extent that game theory is at all comparable to evolution.[1]

The possibility of a direct analogy with a normal poker game has now been eliminated; but an analogy with an imaginable, if unpleasant, kind of poker is still permissible. This game is played with all the rules of ordinary poker but with the bizarre meta-rule that any player that quits the game is killed. This will be called an "existential" poker game. Notice that the same meta-rule could be made with reference to any other game of the kind normally considered by game theoreticians, so that for each game there exists a corresponding existential game.

If we consider all possible existential games, like existential poker, existential chess, and existential checkers, it seems that despite their multiplicity of sets of rules, and despite the differences in strategy between their

[1]If the reader would like to abandon the game analogy completely at this point he may skip to section four without any loss to the empirical presentation, providing that the abandoning of the game analogy is complete and consistent. That is, it would not be legitimate to claim to be abandoning the analogy of a game and then invoke "game-like" payoffs to explain evolution or adaptation. It must be agreed that organisms really are confined to a single world. They cannot be expected, in any way, to increase their stock of anything at all that is not *demonstrably* of use in that world. Any measure or property which can be shown ever to have decreased during the evolution of any phyletic line can never again be used as being identical or correlated with fitness, unless appropriate relations to specific environmental properties are invoked as justification and support.

normal counterparts, there is a unique best strategy for playing any and all existential games. Exactly how this strategy relates to any particular game depends on the rules of the game. Should it be the case that we can specify a biologically feasible pattern of behavior consistent with following this optimal strategy, then we may be in position to build a predictive theory of evolution.

In the absence of any source of variation, the optimal strategy in an existential game is to find some safe move and repeat it at every turn. Consider, for example, two children pushing single kings back and forth on opposite corners of a checkerboard.

If new variance is being introduced with time, repetition of precisely the same move will result in a drunkard's walk to loss. In the children's card game "War," neither player has any flexibility in his moves. Cards are matched and the player showing the higher card takes the lower card. If the matched cards are a pair, five cards are counted out by each player and the fifth cards are matched, the higher card taking all five. A player loses when he has no more cards. Since neither player can alter his play, the game is certain to come to a definite conclusion.

In a poker game, by contrast, a player can modify his behavior in response to both the variation introduced by shuffling the cards and his deviation from some desired state (i.e., by whether he is "winning" or "losing"). It is therefore possible in principle to continue a poker game for ever.

The reason that a poker game can last forever, while a game of "War" cannot, is that homeostatic behavior is permissible in poker but not in "War."

One measure of a player's success at existential poker is the fact of his persistence in the game, but this in itself does not permit ranking of the relative quality of two players except in a retrospective way. A more valuable measure would be the value of the poker chips in front of each player. This is not because the chips can be cashed in, since by definition of the game they cannot, but rather because the size of the stack of chips in front of a player is a partial measure of his ability to withstand disturbance due to future variance in the cards. The measure is only partial, however, since a good player with a relatively small stack of chips may actually have a better probability of persistence than a poor player who has been lucky in the recent past. The true measure of prospective success, i.e., persistence, is the player's ability to behave in a homeostatic fashion so as to avoid or counteract the effects of introduced variance.

In fact, for any existential game with variance, the optimal strategy for any player is to maximize his homeostatic ability, (cf., Slobodkin, 1964). In fact, this constitutes a solution of the game in the game theoretical sense (Anatol Rapoport, personal communication); but it is in a sense obvious and not very helpful unless the technique for maximizing homeostatic ability is more explicitly stated.

In a normal, zero-sum type of game, there is a contest between players for payoff; but if there is no termination to the game, then the player is concerned about his own state, rather than payoff, and all his tactics are played accordingly. An intelligent player in an existential game must assess the probable future significance of each kind and magnitude of variance or perturbation and relegate his resources accordingly. Certain events indicate only minor, rapid fluctuations, and it would not be appropriate to change his style of play drastically in response to them, while others require greater adjustments. The intelligent player must keep in mind how a change in the game is likely to alter his present state, and whether or not such an alteration is desirable. If his present state has not changed for some time, it is probably more desirable to maintain it than change it. If his state is changing, he must take cognizance of the direction of change and whether or not his ability to maintain himself against probable future perturbation is increasing or decreasing. Precisely how this is to be done will vary with the rules of the particular game.

ORGANISMS AS MAXIMIZERS OF HOMEOSTATIC ABILITY

Certain Biological Properties of all Organisms

The applicability of the concept of an existential game to evolution in any interesting way is contingent on the logical possibility of a non-intelligent player or an automation being able to be effective in an existential game, without being explicitly programmed with information about the future. The appropriateness of its activities must be assessed not only in the degree to which it adjusts homeostatically to present perturbation but also the degree to which it maximizes its own capacity to make such adjustments in the future. That is, it must "learn," in some sense, to keep up with a hanging environment and be able to abandon or modify its preconceptions about that environment.

Ashby (1960) has developed an electronic device, the "homeostat," which essentially does this, thereby demonstrating the logical possibility of such a device. It is critical, however, to show that organisms, either singly or in aggregates, are, in fact, devices of this sort.

I do not have to prove the existence of homeostatic mechanisms in organisms. There is an extensive literature documenting their occurrence at all levels from the biochemical through the ecological. I need only be concerned with their relevance to environmental variance and to evolutionary process.

At no point will I postulate any form of precognition, intellectual insight, or group survival mechanism for organisms. I will also refrain from simplifying assumptions and from statements which are applicable to one class of organisms but completely inapplicable to other classes. I will also abstain from assuming the applicability of any particular contrary-to-fact mathematical function or distribution, except in a purely illustrative and non-predictive context. Failure to observe these restrictions is due to error on my part.

I assume the following properties to be essentially valid for all organisms:

(a) While the responses that organisms make to environmental change may begin at essentially the same instant as the change itself, the adjustment process takes time for its completion. Different processes differ in their time requirements. Behavioral responses may be very rapid (fractions of seconds), while disturbances of age frequency distributions may take several generations to correct; and internal reorganizations of genetic material may take longer still. I am referring here both to short-term environmental perturbations as well as permanent or unprecedented changes in the environment.

(b) Any response of an organism or group of organisms to an environmental change involves some commitment of the organisms' resources in a broad sense. With qualifications that would have to be made in individual cases, any activity on the part of an organism will alter, for better or worse, its ability to simultaneously perform other activities. For example, an animal afflicted with almost any disease is typically more susceptible to other diseases. An animal that has just been subjected to an acute physiological stress is relatively less able to withstand further stress (Kinne 1963). Bateson (1963) has constructed a general hypothesis of an "economics" of flexibility-restoring mechanisms, in which slow, deep-seated, physiological changes restore the organism's ability to use its rapid, short-term response mechanisms in a flexible way. I will assume, with him, that there exists a hierarchy of flexibility-restoring mechanisms in organisms. This will be explained more fully below.

(c) In a constant environment, in either the laboratory or the field, any population of organisms will eventually approach a numerical steady state, or fluctuate around some constant mean (cf. the recent discussion between Ehrlich and Birch, 1967; and Slobodkin, Smith, and Hairston, 1967, which may serve as an introduction to the argument that has surrounded this point). The more crowded organisms are, in any given set of environmental circumstances, the poorer will be the physiological condition of each organism on the average. It is this feature of organisms that permits the development of large populations from small inocula. Note that there are special circumstances under which this pattern does not hold over all density ranges, but these "optimal density" situations occur only at relatively low crowding levels (Slobodkin, 1955).

(d) The response to an environmental change may include gene frequency changes in the population and these gene frequency changes may themselves alter the response of the population to further selective pressures. This is the "genetic homeostasis" phenomenon of Lerner (1954). It may be expected, for example, that strong selection for a specific anatomical feature may result in a lowering of overall physiological flexibility or that selection for a specific behavioral feature may be accompanied by morphological changes. It may further be anticipated that gene frequency changes which alter phenotypes will trigger other gene frequency changes, since the new phenotype will have the effect of changing the organisms' relation to the other features of their environment.

In short, I assume the biological validity of the following statements:

1. Biological reactions take time, and different reactions take different lengths of time.
2. Response to an environmental stress alters the response to other environmental stresses.
3. There is a reciprocal relation between the number of organisms in a population and the general well-being of each individual.
4. Changes in population genetics systems may act as selective forces generating further changes in the same systems.

These are not assumptions in the sense of being either false or improbable assertions that make theory construction easier, but rather in the sense that they are introduced as empirical facts rather than as consequences of logical or mathematical development.

The Effect of Various Sorts of Perturbations on Populations of Organisms

Many of the ideas presented here were derived from Bateson (1963). They have been extended to include ecological considerations which Bateson omitted.

If organisms are, in fact, maximizing their homeostatic ability during the course of evolution, it must be the case that they respond to environmental perturbations in such a way that they not only minimize the departure from steady state conditions caused by the perturbation but also maximize their ability to withstand further perturbations. I will now describe, in a general but not simplified way, how organisms do, in fact, respond to perturbations. It will be apparent that the response patterns of organisms are reminiscent of a robot existential game player.

I start with a biological population in an environment that has been reasonably constant for some time. This population is considered to have sufficient homeostatic ability for this environment, since if it did not it would have disappeared some time ago.

In a population of this kind the individual animals can be characterized by some mean level of physiological flexibility. I am not denying that physiological properties change with age, sex, and genotype as well as with immediate environmental conditions. None of these qualifications deny the existence of some mean physiological flexibility in the population. This flexibility can be measured in terms of the magnitude, rate, and kind of environmental perturbations that the organisms can be subjected to and still recover to a greater or lesser degree.

This population will also be characterized by some constant mean population size, which may be either unchanging or cyclic.

The constancy of physiological flexibility and population size are interrelated since constancy of population size implies that, on the average, survivorship l and fecundity m as a function of age x precisely satisfy the equation $\int l_x m_x dx = 1$. Any change in physiological condition will be expected to alter l_x, m_x, or both. It therefore follows that only if the age specific physiological state of the animals fails to show a temporal trend can the number of animals in the population remain constant.

It should be noted that the relation is not a reciprocal one. That is, constancy of population size does not imply constancy of physiological condition. Consider, for example, two environmental properties, one of which is altered in such a way as to increase population size, if this were the only environmental change, while the other is altered so as to decrease population size if it were the only change. These two properties can now be simultaneously altered so that population size is not changed, but the physiological condition of the individuals is changed. That is, both l_x and m_x are changed, but the integral over their products remains unity. This has been demonstrated empirically by Smith (1963).

While physiological condition and abundance may be constant in a population that has been in a constant environment for some time, I expect that slow gene frequency changes will be occurring because of mutations and genetic homeostasis effects in the sense of Lerner (1954).

If the environment changes, the properties of the population will change. The kind of change will depend on the kind of environmental perturbation. I will consider two extreme types of perturbations and demonstrate that we can predict, in a general way, how populations will respond to them. More precise predictions would be possible, given further knowledge of either the history of previous perturbations or detailed biological information about the organisms in the population. Most significantly, it will be seen that the response of real populations to perturbation is not only homeostatic but also tends to maintain homeostatic ability.

The two extreme types of perturbations, for my purpose, are those which are, in some sense, perceived by the organisms and elicit a series of behavioral and physiological responses and those which are catastrophic in the sense that they are either not perceived or occur so rapidly and with such great force that no active response on the part of the organisms is possible. The kind of distinction I am making is that between the effect on krill of slight salinity change, on the one hand, and being swallowed by a whale, on the other. Most environmental perturbations are intermediate in character, and how a particular organism responds to a particular perturbation depends to some degree on the properties of the organism.

My primary concern will be with perturbations that immediately decrease the organism's ability to withstand other perturbations.

The response of each individual organism to a non-catastrophic perturbation is essentially as indicated in Figure 1.

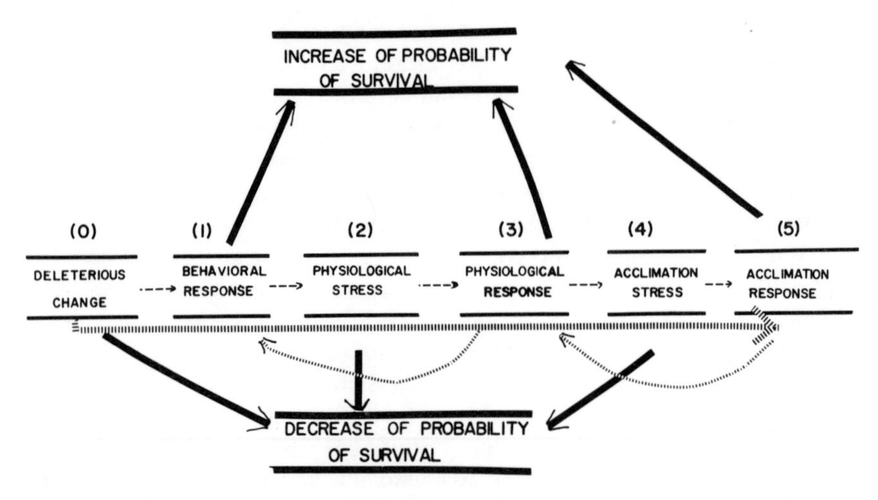

Figure 1. Response of an individual organism to a non-catastrophic perturbation from environmental sources.

The initial response to any perturbation is behavioral. If the behavior in effect eliminates the perturbation, or if the perturbation is reversed sufficiently quickly, no further response occurs. Consider, for example, a dog resting on a floor which is heated locally. The dog is uncomfortable, moves away, lies down again and, in effect, the perturbation has vanished, just as if the floor had been quickly cooled back to its original temperature.

Even if the perturbation is of an absolutely new kind in the animals' experience, say a new chemical or new kind of radiation, the initial response to it will be behavioral, since it will be coded as if it were one of the normal perturbations to which the animals can respond. The response need not be appropriate to the perturbation.

If the response is inappropriate, or if the perturbation is such that no behavioral response will suffice to nullify it, and if the perturbation is a persistent change, then physiological responses will become increasingly significant. Consider that the floor available to the dog is uniformly heated, so that moving away can't help. The dog will become increasingly uncomfortable and will begin to pant.

Further temporal persistence of the perturbation will result in continuation of the physiological response; but to the degree that the state of making the response differs from the normal state of the animal, more and more deep physiological adjustments will have to be made. A sufficiently long period of disturbance will result either in the death of the animal or in physiological acclimation or adaptation. Functionally, adaptation can be thought of as altering the capacity to respond, rather than constituting a response. That is, an animal adapted to a particular situation can be distinguished from a genetically identical animal that has not adapted by differences in their performance under further stress. Bateson (1963) uses the example of a man moved from sea level to a high altitude: ". . . he may begin to pant and his heart may race. But these first changes are swiftly reversible: if he descends the same day, they will disappear immediately. If, however, he remains at the high altitude, a second line of defense appears. He will become slowly acclimated as a result of complex physiological changes. His heart will cease to race, and he will no longer pant unless he undertakes some special exertion. . . . From the point of view of an economics of flexibility, the first effect of high altitude is to reduce the organism to a limited set of states . . . the man can still survive but only as a comparatively inflexible creature. . . . After the man is acclimated he can use his panting mechanism to adjust to *other* emergencies which might otherwise be lethal."

At least with respect to temperature, adaptation and acclimation may adjust either the capacity or the resistance of the organisms (Kinne, 1963). Capacity changes involve alteration in metabolic rates and in activity, while resistance changes involve changes in thermal death points. These properties may be related, and the distinctions between capacity and resistance are possibly also relevant to other sources of stress in addition to temperature.

The response of individual organisms to environmental change will be to simultaneously initiate behavioral and physiological adjustments and to start to change the state of physiological adaptation. If the environmental change is rapidly reversed, either by the action of the organism itself or by an external agency, there will be relatively little change in the deeper physiological properties of the organism. To the degree that the environmental change is greater in magnitude or duration, deeper changes will occur. The fact that the more shallow responses are the more rapid permits the organism to act as if it were correctly assessing the expected permanence and importance of environmental perturbations. That is, if a particular environmental change could be clearly assigned to the category of permanent important changes, then an appropriate, deep, long-term response could be made immediately. Lacking this precognition, individual organisms respond to any change by immediately beginning a series of changes on all levels. The perturbation only seriously effects the deeper, more slowly responding, mechanisms if it is, in fact, relatively permanent.

While, to a degree, behavior acts as a bulwark against the necessity for physiological change, the simplicity of the hierarchical relation should not be overemphasized. It is probably valid, however, to assert that organisms respond to rapidly fluctuating features of their environment behaviorally and to slower fluctuations physiologically, and that just as learning restores flexibility to behavior, so does adaptation restore flexibility to physiological responses.

Therefore, individual organisms act, to some degree, in the way that we would expect of a robot existential game player. However, a player of an existential game in an environment that shows no temporal trends had best remain in an unchanging steady state. No single organism, with the possible exception of some flatworms, protozoans, and coelenterates seems capable of doing this. The capacity of animals to respond to their environment varies with age. There is not, however, any similar necessity for the properties of a population of organisms to change with time.

The effect of slow, moderately deleterious, perturbations on populations of organisms is shown in Figure 2a.

The immediate effect of the perturbation is to lower the survival probability and homeostatic ability of the individual organisms (as discussed above). This results in a relative increase of mortality and decrease of fecundity. But the constancy of population size is contingent on the relative constancy of fecundity and mortality so that there is decline in population size associated with the perturbation. It is also to be expected that genotypic differences between animals will, to some degree, determine which live and which die. To the degree that this is so, the perturbation will accelerate the rate of gene frequency change.

These three effects of perturbation are, however, interrelated in an interesting way.

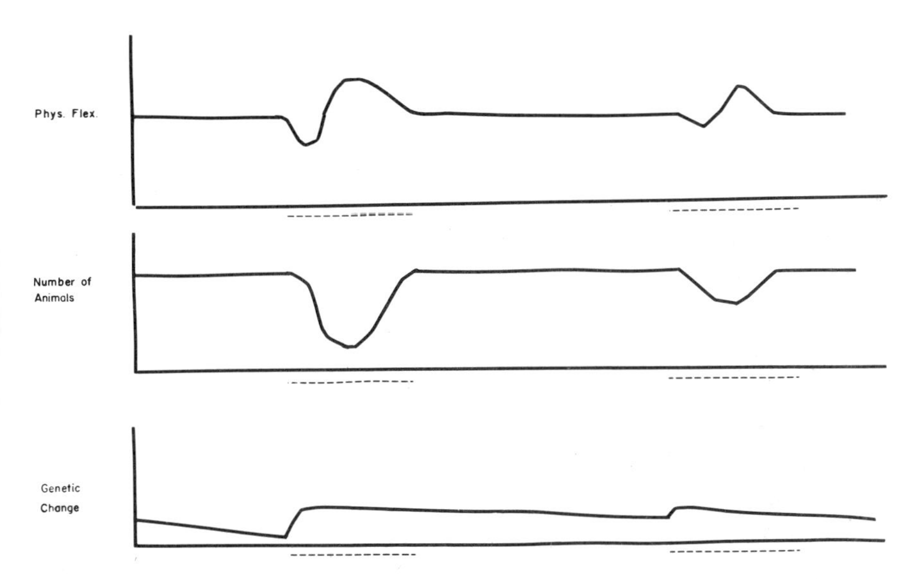

Figure 2a. Changes in physiological flexibility, population size, and gene frequency in populations subject to a non-catastrophic perturbation. Light dotted line indicated the period of perturbation.

The increase of relative mortality lowers population size; but the lowering of population size will, in turn, result in an improvement of the physiological flexibility of the survivors. The mechanism involved derives from the fact that crowding has a generally deleterious effect on the individuals of a population. Alleviation of crowding, conversely, improves the physiological condition of the survivors to some degree. The precise extent of this effect will vary from species to species, but it has been observed in all laboratory populations (Slobodkin, 1961, Ehrlich and Birch 1967) and is almost universal in nature (Slobodkin, Smith, and Hairston, 1967).

If the environmental disturbance is not too severe, the improvement of physiological condition associated with reduced crowding will result in a cessation of the population decrease and the attainment of a new steady state, despite the persistence of the environmental disturbance. How many animals will be in the new population and their specific physiological properties will vary with the animals and with the perturbations. In particular, the physiological benefits associated with decreased crowding need not be intimately associated with the perturbation that produced the decreased crowding. For example, a poison that kills off some animals in a food-limited population may, by increasing the food supply for the survivors, permit them to better withstand the poison's effects.

To the degree that environmental perturbations are successfully countered by behavioral and physiological mechanisms, and to the extent that the animals in the population are genetically similar, perturbations will leave no permanent genetic trace on populations. Genetic change will only occur if there exists differential mortality or fecundity among the genetically different members of the population, or due to mutation or drift. I will, as a matter of convenience, ignore both mutation and drift. Their inclusion would not materially alter the present argument.

If the environmental perturbation is of relatively short duration, the population may be expected to return to essentially its initial condition of abundance and physiological flexibility after it has passed. If an identical perturbation reoccurs relatively soon after the first one, the ability of the population to respond by physiological adjustment of the individual members should be somewhat enhanced, as a result of the gene frequency changes associated with the initial occurence? There should, therefore, be relatively lower mortality and relatively less genetic change associated with the reocurrence of a perturbation than with its initial occurrence.

In short, any perturbation of the environment to which some physiological response is possible, if it occurs with sufficiently high frequency, will eventually come to be dealt with by a behavioral or physiological mechanism and will cause a minimum of genetic effects on its later reoccurrences. Conversely, changes in gene frequency in a population represent a response to some disturbance for which no adequate behavioral or physiological adjustment mechanisms exist.

It is possible to imagine, however, perturbations which are so sudden and so severe that no appropriate response on the part of the organisms is possible, i.e., catastrophes. Such perturbations result in immediate death of some animals, but, in the most extreme cases, no specific selection for ability to withstand the catastrophe by physiological or behavioral adjustments. The consequences of extreme catastrophic perturbations are shown in Figure 2b.

The catastrophic elimination of animals, without regard to genotype, will have the effect of selecting for the ability to increase rapidly, without regard to the specific source of the catastrophe that initially depleted the population. That is, if a slow change in salinity, for example, was the initial perturbation, those animals with a more appropriate salinity tolerance and, simultaneously, with the ability to reproduce rapidly in response to the absence of salinity-sensitive individuals, will be selected for. In the response to a pure catastrophe, on the other hand, in which there is no mechanism of escape or amelioration other than chance, the only effective selection is for rapidity of increase after the event.

It is well established that there is an inverse relation between adult body size and the ability to increase rapidly (Bonner, 1965). One general re-

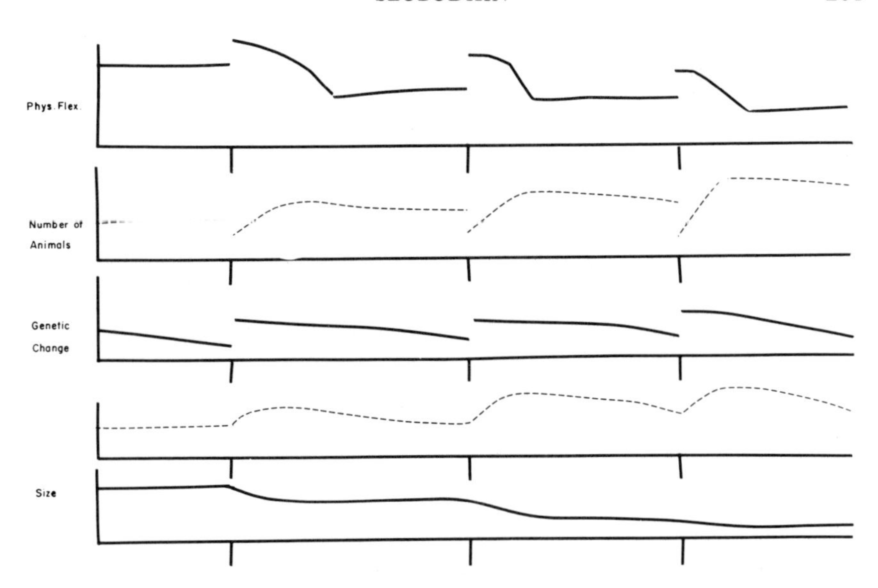

Figure 2b. Changes in physiological flexibility, population size, gene frequencies, and individual body size following a catastrophic perturbation of the environment.

sponse that may be expected in the wake of catastrophic perturbations is the evolution of small body size. Recent evidence indicates that this phenomena extends even into the molecular range (Mills, *et. al.*, 1967).

It is also the case that smaller animals tend to have lower overall flexibility to many kinds of perturbations than do large ones, and that smaller animals can maintain a numerically greater population than larger ones on a given resource supply. These effects, in turn, increase the susceptibility of populations of small animals to perturbation and tend to enhance the catastrophic aspects of whatever large perturbations may occur. Eventually the physiological flexibility will be so reduced that minor disturbances will take on the aspect of the type of perturbation discussed in connection with Figure 2a.

In short, individual animals respond to perturbation by simultaneously starting behavioral and physiological mechanisms. Should the perturbation prove to be a minor or short-term event, the shorter term response mechanisms will deal with it, leaving the deeper, slower response mechanisms unchanged. To the degree that the perturbation is of greater magnitude or duration, deeper and deeper responses are called forth from the individual. To the degree that these mechanisms are inadequate, more profound changes occur in the population. The response of a population to a completely novel, major, perturbation is a part genetic; but if this perturbation proves to be periodic, and if the period is sufficiently short, the response to its re-occurrences will be physiological or behavioral rather than genetic.

If a population is highly adapted to its environment, the probability of events in that environment will be highly correlated with the probability of the population withstanding that event, without loss of homeostatic ability. To the degree that this does not hold for any particular event, the population will change. This change will not be a general improvement in any sense but rather an adjustment to the specific disturbing event, presumably at the cost of some other kind of homeostatic ability.

The response of a population of organisms to any set of perturbations is, therefore, almost precisely what would be expected of a robot player of an existential game, and does in fact conserve the homeostatic ability of the population to face further perturbations.

CONCLUSIONS

I have attempted to state, in fair detail, how behavioral, physiological, ecological, and genetic mechanisms interact with each other and with the environment to maximize and conserve the homeostatic ability of populations of organisms. Each of the mechanisms discussed is well documented, and no simplifications have been introduced in the discussion. I suggest that the maximization of the ability to behave homeostatically (using behavior in a broad sense) in their particular environment, is the only thing that has always failed to decrease in the process of evolution. If this is so, it has as a consequence the assertion that no single measure can be considered as a valid indicator of an organism's evolutionary success or evolutionary promise, except to the degree that it is correlated with homeostatic ability in some particular environmental situation.

This in no sense denies the fact that any gene which permits an organism to contribute more than its share to subsequent generations will have a selective advantage. Other things being equal, higher reproductive rates are selected for. Unfortunately other things are almost never equal. The physiological and behavioral processes which would result in higher reproductive rate in one set of circumstances will decrease reproductive rate in another. This is amply documented by the evolution of several phyletic lines, including the primates, in which the relation to environmental circumstances has been such that lower, rather than higher, fecundity has resulted in higher effective contribution to subsequent generations. Lack (1954) and others have amply demonstrated the absence of a positive relation between higher than normal fecundity and higher effective reproduction.

Similar remarks apply to abundance. There is no evidence that rarity *per se* increases the probability of species or population extinction. Granted that a species becomes rare during the process of becoming extinct, and the last survivor is rare indeed; but this does not imply that rarity is the cause of its extinction. There are, in fact, peculiar relations to the environment which may simultaneously produce rarity or extreme restriction of range and at the same time may involve a high susceptibility to environmental change. There are also counterexamples in which rare species have outlasted much more abundant species (compare, for example, passenger pigeons and Kirtland's warblers). Further, it has been shown experimentally that as the culture conditions for laboratory populations of Hydra approach those in nature, the number of animals in the steady state populations is reduced by a factor of five to eight (Lomnicki and Slobodkin, 1966; Ritte, 1968). That is, abundance may, on occasion, be a pathological response quite unrelated to natural circumstances.

Therefore, to assume any specific property of a population or organism as being a valid predictive measure of fitness for that population or organism, without explicit demonstration of how that property relates to the animals' environment and to the other properties of the organism which also relate to that same environment, is a projection of an external set of values onto the evolutionary process. It is no more legitimate than the nineteenth-century projections of moral and ethical qualities onto evolution.

Any theory with pretensions to predictive power, which is built on a measurement which is assumed correlated with fitness, but which does not take into account the specific environmental problems faced by the organisms, is no stronger than its initial measurement, regardless of its mathematical elegance or complexity. This does not require a cataloging of environmental properties for the world's two million-plus species before one can build an evolutionary theory with predictive power. Fortunately, there are broad commonalities between classes of perturbations and classes of response mechanisms, many of which are adequately documented in textbooks of ecology and comparative physiology.

The development of a predictive theory ultimately depends on being able to specify when a population is in a better or worse evolutionary state. For this purpose an objective definition of adaptedness (Dobzhansky, 1967) is necessary. I tentatively suggest that such a definition could be stated in terms of the probability of occurrence of particular environmental states and the probability of a population surviving each such state. In general, a well-adapted population is one in which the probability of surviving a particular environmental state is greater the greater the probability of encountering that state. Any departure from a perfect rank order correlation between the occurrence of environmental states and the probability of surviving would indicate room for improvement in the state of adaptedness.

If certain simplifying assumptions are made, this concept of adaptedness can generate more formal theorems. For example, if the environmental states are finite in number and independent in temporal sequence, then there is a straight line relation between environmental state probability and survival probability, except for highly improbable states for which survival probability will be excessively low. I have attempted to avoid this type of simplification in this discussion, and will not pursue this path further.

Given an adapted population, should either the environment or the organisms change without appropriate reciprocity, the level of adaptedness will decline. Whether the changes in the organisms had best be behavioral, physiological, or ecological is determinable from the spatial and temporal properties of the environmental change. To merely tell the animals to reproduce more and live longer is cold comfort in the face of actual problems of adaptedness.

In what sense have we produced at least a skeleton of a predictive theory? We have seen that certain classes of response to environmental perturbations cannot occur, given the properties of organisms and the existential nature of the evolutionary process. More specifically, the arguments presented above permit certain general predictions to be made about how the evolutionary process occurs. For example:

1. In well-adapted organisms there is a strong correlation between the probability of environmental events occurring and the probability or responding to these events without any genetic change. That is, gene frequency changes are a type of last resort in the process of adjusting to environmental change.

2. Extinction of populations and species is not correlated with rarity per se.

3. Periods of high rates of extinction are also periods of high rates of gene frequency change.

4. Consistent prediction will not arise from evolutionary models in which either abundance or absolute reproductive rates are taken as direct measures of probable evolutionary success.

LITERATURE CITED

Ashby, William Ross, 1960, *Design for a Brain*, 2nd. ed., Wiley, New York.

Bateson, Gregory, 1963, The Role of Somatic Change in Evolution," *Evolution*, 17: 529–539.

Bonner, John Tyler, 1965, *Size and Cycle: an Essay on the Structure of Biology*. Princeton University Press, Princeton, N.J.

Dobzhansky, T., This volume.

Ehrlich, P. R. and L. C. Birch, 1967, "The Balance of Nature and Population Control," *Amer. Natur.*, 101:97–108.

Huizenga, Johan, 1949, *Homo ludens; a Study of the Play-element in Culture*, Translated by R. F. C. Hull, Routledge, and K. Paul, London.

Kinne, O., 1964, "The Effects of Temperature and Salinity on Marine and Brackish Water Animals. I. Temperature," *Oceanogr. Mar. Biol. Ann. Rev.*, 1:301–340.

Lack, David Lambert, 1954, *The Natural Regulation of Animal Numbers*. Clarendon Press, Oxford. VIII.

Lerner, Isador Michael, 1954, *Genetic Homeostasis*, Oliver and Boyd, Edinburgh.

Lewontin, R. C., 1961, "Evolution and the Theory of Games." *J. Theoret Biol.*, 1:382–403.

Lomnicki, Adam, and L. B. Slobodkin, 1966, "Floating in *Hydra littoralis*," *Ecology*, 47:881–889.

Medawar, Peter Brian, 1960, *The Future of Man*, Methuen, London.

Mills, D. R., R. L. Peterson, and S. Spiegelman, 1967, "An Extracellular Darwinian Experiment with a Self-duplicating Nucleic Acid Molecule," *Proc. Nat. Acad. Sci. U.S.A.*, 58: 217–224.

de Montaigne, Michel, "Of trivial subtleties," In *The Essays of Michel de Montaigne*. Jacob Zeitlin, ed. and trans., Alfred A. Kopf, Inc., Vol. I, pp. 273–275 (1934).

Rapoport, Anatol, 1960, *Fights, Games, and Debates*. Univ. of Michigan Press. XVI.

Slobodkin, L. B., 1955, "Conditions for population equilibrium," *Ecology*, 36:530–533.

______ 1961, *Growth and Regulation of Animal Populations*, Holt, Rinehart, and Winston, New York.

______, 1964, "The strategy of evolution," *Amer. Sci.*, 52:342–357.

Slobodkin, L. B., F. E. Smith, and N. G. Hairston, 1967, "Regulation in Terrestrial Ecosystems, and the Implied Balance of Nature," *Amer. Natur.*, 010:109–124.

Smith, Frederick E., 1963, "Population Dynamics in *Daphnia magnum*, and a New Model for Population Growth," *Ecology*, 44:651–663.

Warburton, F. E., 1967, "A model of Natural Selection Based on a Theory of Guessing Games," *J. Theoret. Biol.*, 16:78–96.